Motorbooks International

FARM TRACTOR COLOR HISTORY

COMBINES & HARVESTERS

Hans Halberstadt

❧

Dedication

To Cliff and Onalee Koster, California farmers of the old school—vine ripened and Grade A Prime.

Acknowledgments

I am particularly grateful for the help of East Bay Regional Park District's Ardenwood Historic Farm and the generous support of David Cook.

Cliff and Onalee Koster; J. I. Case Company, particularly Gerry Salzman and archivist David Rogers for extremely generous support with information and images.

California Planting Cotton Seed Distributors; University of California Davis/F. Hal Higgins Collection; Midwest Old Time Thresher's Reunion, Mt. Pleasant, Iowa; Lorry Dunning; Paul Reno; Holt Brothers, Stockton, California: although the old Holt combine company got sucked up into the massive yellow Caterpillar, their legend and lore live on. Many thanks to Vic Wykoff, Kimberly Christolos, and all the other Holt folks who saved the Model 32 and who brought it back to life behind a team just for this book.

Dr. John Hope, Esq., a member of the advisory staff and a superb scout; Cornie Boersma, a Class A mule skinner from Spearfish, South Dakota ; Jerry Galliher, Belle Fourche, South Dakota; Alan Harper, Sundance, Wyoming; Robert Genet, for photographic assistance at the Antique Gas and Steam Engine Museum, Vista, California; Barbara Bywater, AGCO, Independence, Missouri

All this would not be possible without the efforts of my Motorbooks teamplayers: Editors Greg Field and Jane Mausser and Designer Amy Huberty.

First published in 1994 by Motorbooks International Publishers & Wholesalers, PO Box 2, 729 Prospect Avenue, Osceola, WI 54020 USA

Library of Congress Cataloging-in-Publication Data

Halberstadt, Hans.
    Combines and harvesters /
    Hans Halberstadt.
        p. cm. — (Motorbooks International farm tractor color history)
    Includes index.
    ISBN 0-87938-944-3
    1. Combines (Agricultural machinery)—History. 2. Harvesting machinery—History. 3. Combines (Agricultural machinery)—United States—History. 4. Harvesting machinery—United States—History.
    I. Title. II. Series.
    S696.H34 1994   94-32033
    681'.7631—dc20

On the front cover: This International Harvester 320 single row cotton picker was part of the first generation of really practical cotton harvesters.

On the back cover: A Holt Model 32 harvester restored to virtually new, working condition; John Tower and daughter bringing in the wheat with the old John Deere Model 55.

On the frontispiece: Detail of a Circa 1900 Sandwich "Jupiter" corn sheller.

On the title pages: A flotilla of modern combines sweep across the grain fields of the American midwest.

Printed and bound in Hong Kong

Contents

Preface

Antique farm equipment has become something of a modern mania. Dozens of shows around Canada and the United States celebrate the rich, fascinating heritage of our farming forefathers—and foremothers, too. Thousands of people come to see the glittering old "Johnny Poppers" and the huge, lumbering Case, Advance, or Rumley steam tractors chuff their stuff. You can get more money from a collector for that rusting hulk out back of the silo than you had to pay for it new, back when it was still shiny and could still run.

When somebody trots out their handsome matched team of Belgians or Shires to demonstrate how we used to plow, harrow, mow, reap, and thresh, I'm always awed by what a beautiful, elegant, fascinating set of

technologies and systems were available fifty or a hundred years ago. I have to remind myself—it wasn't always romantic at the time: the horses often spooked and sometimes ran away with the binder or the wagon, the work was dirty, long, dangerous . . . and yet, there was something special and romantic about farming and farm equipment back then. You can still see it in the portrait photos the farmers had made at threshing time, the shot of the combine with the forty-four horse hitch, back in '27, or '02. You see it in the advertisements and brochures from the 1930s, in the marvelous farm newspapers from the 1880s, and you can see it in the cast iron and forged steel and painted wood of the machines that still survive. These folks were proud of what they were doing, in the

Opposite, while Massey-Harris tractors of this vintage are prime collector material, this first-generation self-propelled combine will probably never be restored; like others of its breed and generation, it is out to pasture and allowed to decay. When it was new, in 1953, the advertising copy for it read, "NEW 90—*The biggest capacity combine on wheels! Here in the 90 Self-Propelled you have an example of the "look-ahead" policy at Massey-Harris. As a dealer, you can be part of it . . . know its value when you're out on a demonstration or are up against tough competition. Make It a Massey-Harris—America's fastest growing full line implement company!"*

International Harvester 141-SP; this machine came out of the factory between 1954 and '57, and was the first from International Harvester to offer an optional corn head, or headers from 10ft to 14ft wide. Its six-cylinder engine is probably rusted solid now, but it once provided more than adequate power for the nearly 9,000lb machine.

way the machines were made and the way they were used, and it still shows.

These old machines remind me of the way modern jet fighter aircraft are designed and built today; you can see in every part, every rivet and turbine blade, a commitment to sturdy, sleek, perfection in every tiny hidden detail. Both kinds of machines were designed and built by people who knew they

Above, here's one of the very few really restored Circa 1900 Sandwich "Jupiter" corn shellers you are likely to encounter anywhere, and this one still works the way it was intended—with the help of horses. It provides some of the entertainment at Midwest Old Threshers Reunion.

were building something important, a tool that had to be efficient and economical . . . a weapon, in both cases, in the war for survival.

If you attend one of these shows—and you should—you'll see row upon row of beautiful old Case, John Deere, Rumley, Advance, and other tractors glittering in their perfect, showroom fresh paint and authentic decals, but there will be few if any old combines, mowers, reapers, or harvesters. Sometimes an elderly thresher, restored or rotten, will be rolled out and belted up for a demonstration, but not often. The tractors are the stars of these shows, and with good reason. You can hardly drive a combine in a parade.

Left, here's where old combines go to die—the bone-yard at the University of California Davis campus, the ag school for the California university system. Although this Holt probably won't be resurrected, many important examples of agriculture technology have been brought back to life from here by UCD students, under the guidance of Lorry Dunning—the "Mr. Fixit" of California farming history.

Below, some original paint still protects the woodwork of a sadly decaying mammoth Holt combine at UCD, the ghost of what was once a state-of-the-art harvester.

Most of the old combines and harvesters you see on display are parked out in the corner of somebody's pasture or at the edge of the field where they died. You see them everywhere around the US and Canada, weeds and small trees growing up around them, the elements slowly eroding the paint, steel, rubber, and wood back to the earth. It is odd, and sad, that these big Minneapolis-Molines from the 1950s or the International Harvesters from the '60s that were fabulously expensive, state-of-the-art technologies and a farmer's pride and joy, are discarded so quickly. For most, a working life is about ten seasons—around 2,000 hours of operation before they are parked forever.

An old tractor of any make, in any condition, will normally command serious money to change hands—but you can have your pick of old combines if you have a big enough trailer and will come and fetch the thing out of the field. Unlike old tractors, old combines take up a lot of space, are complicated and difficult to restore, and about the only place you can show off one is in a quarter section of ripe wheat or barley.

So the machines in this book aren't the glittering restored collectibles that you'll find in the tractor books. Few people collect them and fewer restore them. The old ones I have included either work for a living or are recently retired, with a very few exceptions. Most still work for a living, and will until

their bearings completely wear out, and then they, too, will get parked out of the way to rust in peace. But there are some folks (like my friend Cliff Koster who keeps an ancient Harris harvester in the barn, along with the harness for the mules who pulled it) who preserve the old machines out of affection for the machines, and with a long view and a wide perspective for their role in North American agriculture. There

are also, increasingly, many places like Ardenwood Historic Farm that are designed to preserve, protect, and defend our agricultural tradition.

Finally, this is a book about both grain combines and other harvesting machines. That includes a pretty broad spectrum of machinery: cotton pickers, hay mowers and rakes, cranberry pluckers, tomato harvesters, and a few other exotic old clankers.

Valley of the Heart's Delight

You'd never know it, but the town where I now live was built on what was once some of the finest, most productive farmland in America. Well, actually, it was in Spanish California back then, two hundred years ago, when the old padres and their Indian converts began the process of developing the virgin landscape to farmland. The topsoil was deep and rich, with abundant artesian wells providing easy irrigation. That combination, plus the long growing season, meant that you could just about throw a seed at the ground and it would grow. If you could shoo away the elk and deer that roamed this valley, pretty soon you'd have a crop. It was, even 200 years ago, a bountiful land.

Using the most primitive cultivation methods, the most ancient harvesting techniques, the old missions were awash in grain. There was more than enough for bread, for the horses, for export—had there been a way to get it to market, but that would have to wait. The whole economy of California, from 1777 when my home town,

Opposite, when the binder is working well, the sheaves are nice and neat, just like these. They will be allowed to cure in the sun and wind for a few days before being taken in, out of the field and the risk of rain, to be threshed or stored for threshing later in the winter.

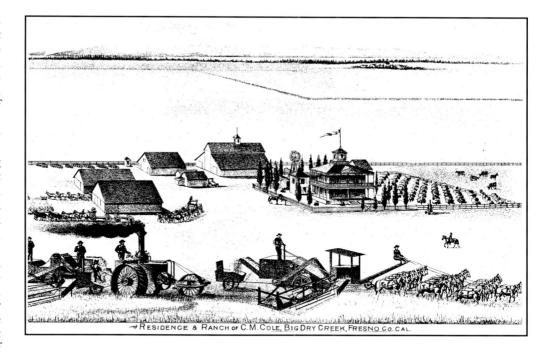

RESIDENCE & RANCH OF C.M. COLE, BIG DRY CREEK, FRESNO CO. CAL.

When they figured out how to farm the great Central Valley of California, just after the Civil War, the state almost instantly became the leading wheat producer of the US. Huge farms, like C.M. Cole's "Big Dry Creek" ranch near the sun-blasted town of Fresno produced oceans of wheat, barley, and oats from the virgin soil. This Thompson & West lithograph commemorates a harvest around 1880, the first year the state was #1 in grain production, and shows two methods of combining, one with a twenty-horse hitch, the other with an approximately 20hp steamer. On the small farms of Maine, Virginia, and Pennsylvania, the cradle and the flail were still in use, and the simple "groundhog" thresher was still considered modern machinery.

NEW FOR 1842!!! The new, improved Ground Hog threshing machine from J.I. Case! Now with all wood cylinder, and advanced, wood tooth design! Modern farmers from Pokipski to Peoria are saying, "shore beats thrashing the old way with the flail!" and you will too. Will thrash up to 200 bushels a day! Available with advanced horse treadmill power, or with economical, hand cranking! Get yours today! (With apologies to) J.I. Case

Who is that masked man? It's Ira Bletz, the curator of Ardenwood Historic Farm who is entirely

too clean to be authentic, but give him time and a few hundred more bundles.

San Jose, was founded, until 1849 and the Gold Rush, was based on sprawling cattle ranches and the export of cattle hides, the only commodity that could be shipped to market without spoilage.

The land lay mostly fallow until American statehood in 1849. When California was flooded with humanity during the gold rush, it was a hungry humanity, and some of the smart immigrants made their gold by feeding the multitude. For a while, in San Francisco and in the gold fields, a barrel of flower cost $100 in gold coin, a bag of rice $50, an apple cost $5, and a dozen eggs might be worth more than their weight in gold. A lot of unsuccessful miners decided to try a different kind of prospecting, and little farms sprouted anywhere there was water for irrigation and where the Mexican land grant holder wasn't standing guard with a gun.

They discovered that just about anything grew here, and soon enough just about everything did: wheat, barley, rice, berries, grapes, tree fruits, lettuce, leeks, garlic and onions, corn, beans, forage crops of all kinds. And as soon as they could, these products were sent to market—fresh, or in the form of processed food and drink. My house stands on ground that grew grapes that made wine and brandy 130 years back, and up the road folks were growing plums and apricots, wheat and hops. Although all that farmland is now just about entirely encrusted with homes and businesses, the history of my community—like the history of the whole American community—is based on the abundance of the land and on farms and farmers. Until about 1960 farming was the principal industry here; then, in the Spring when the blossoms of hundreds of thousands of fruit trees filled our valley, they used to call this place "The Valley of the Heart's Delight."

Well, times have changed and the orchards are almost all gone here, but farming is still the first business of California and America. It is a different kind of farming now, a better kind of farming in many ways. Food products are cheaper, more abundant, and are available in greater variety than ever. Farming is more efficient than ever, bringing healthy food to more people than even

in those bygone days of the family farm.

That abundance and efficiency is the result of innovation and mechanization. One hundred and fifty years ago it was easy to grow far more wheat, for example, than you could harvest. The process of harvesting any crop was slow, hard, painful work. It was often wasteful as well. Here in California, where tillable land was more abundant than labor, that meant that a huge proportion of every crop rotted on the ground or on the tree or vine because there was no one to harvest it.

The keys to American prosperity have always been recognized as the ability to produce agricultural commodities and the ability to transport the harvest to market, in abundance and at low cost. Both of these

things were accomplished—by the tractor, the combine, and the railroad—by a happy coincidence of the industrial revolution, the availability of vast fertile land, and a flood of intelligent, energetic immigrants from Europe, all during the latter half of the 19th century.

The mechanized harvest is a rather new thing, a process of technological development that is still not complete. In the lettuce and broccoli fields near here, the harvest is still brought in by men and women slowly working their way down the rows, bent over, knife in hand. That's the way just about everything was harvested until about only 100 years ago, when the first really efficient combines and harvesters started to go into the fields in sig-

When Tom Coles retired a few years ago he didn't know anything about horses or driving a team—and there are still times when the horses try to tell him he still has a lot to learn. But he's one of the folks who have found pleasure in doing things the old way, and he's learned to do them well.

nificant numbers. The effect was immediate and profound. As the farm was mechanized, people were freed of much of the drudgery and uncertainty of life. Although we tend to romanticize our early agricultural heritage, it was often a very grim experience.

Hans Halberstadt

13

Chapter 1

Grain Harvest Heritage—A Short History

IN THE FARM SHOP

Grain has been cultivated and harvested for at least 5,000 years. It was and still is a chancy business; if you harvest too early, the heads won't be fully formed or ripe, but if fully ripe the heads will easily "shatter" and spill the grain on the ground before it can be collected. Traditional harvesting methods are so inefficient that poor people were able to survive by "gleaning" from the ground the 10 to 30 percent of the grain that was spilled during harvest.

At first the ripe heads were cut from the stalks with knives made from chipped stone, then from copper and bronze. The amount of labor involved in this kind of harvest was tremendous, up to about 300 hours to reap just one acre. About 5,000 BC the Egyptians reshaped the knife, adding a deep curved pocket that trapped the straw, and the sickle was invented. With this tool a harvester

There are plenty of folks still alive who remember when the cradle was considered a practical, appropriate harvesting tool for cutting grain, and this is the basic tool that cut the grain that fed America and Canada until about 1875. If you have the strength and the skill for it, you can cut an acre of grain a day with one, but then somebody else has to come along and tie it up in sheaves, collect the sheaves into shocks, and then—after a week or so in the sun—collect them all to be threshed in the barn.

might cut as much as one acre in a single long summer day.

Later, in quite recent times, the blade and the handle of the sickle were lengthened to form the two-handed tool we call the scythe. That conversion allowed the harvester to finally stand upright. That development happened spontaneously in several places in Europe during the 18th century, and was perfected in colonial America around the time of the Revolution with the addition of slender wooden "fingers" above the blade. With this new tool, between three and five acres could be cut in a day. Some of the grain heads will shatter and be lost, but this tool and technique was still in common use in the United States well into this century, right up to World War II. And until about 1880 this was pretty much the state of the art of grain harvesting for the US, Canada, and throughout Europe.

Cutting the grain was, in many ways, the easy part; getting it out of the field and off of the straw was work. While men (who were called "reapers") traditionally did the cutting, women and children (who were called "bandsters") followed behind, gathering up the cut grain in armfuls called "gavels" and tying each shock with two small hands full of straw. These shocks were then gathered up and carried (by hand or cart) into barns, away from the weather. The

grain had to be dry enough to prevent spoilage, but not so dry that all the heads shattered before they were threshed.

Normally threshing—the process of separating the kernels of grain from the straw—waited until winter. Then, when there wasn't a lot to do and when the air was often crisp and dry, the farmer and his family and his serfs could attend to the slow, dreary business of "thrashing" out the grain. And it really was a literal process of thrashing the grain; despite the spelling of the word, nearly all American farmers still pronounce the word thresh as "thrash."

The stocks were typically spread out on the floor of the barn or, if the farm had one, a dedicated building called a granary. Livestock were often used to break out the grain by being driven around in circles on the thrashing floor, but they had a tendency to add their own unsanitary contributions to the grain. An alternate method used a "flail," a simple club that was used to manually beat the heads of grain; this was a notoriously man-killing job, but it was either flail or starve for centuries.

But that wasn't the end of the process. The grain still had to be winnowed—separated from the chaff, a procedure usually done with the help of the wind. Finally, you had something to bake bread with. But it took a tremendous amount of time and effort to cultivate, reap, thrash, and winnow the

RUTH gleaning in the fields of Boaz, obtaining barely enough Grain for individual necessities.

THE CROOKED SICKLE. By systematic and hand labor small quantities of Grain were harvested by working from dawn to dark

THE CRADLE. By which the farmer was enabled to raise small quantities of Grain for market.

A REAPING MACHINE, crude, rough, heavy and incomplete, but the dawning of a New Era in raising Grain.

The Johnston Harvester Company's Wrought Iron Harvester.

The Johnston Harvester Company's Self Raking Single Reaper.

The Johnston Harvester Company's Self Raking Reaper and Mower.

Above, the history of harvesting grain, according to The Johnston Harvester Company. *Asher and Adams' Pictorial Album of American Industry 1876*

Right, there was a time, believe it or not, when the idea of a seat on a mower was a very controversial topic in American agriculture. There was a widespread belief that any farmer who would ride, rather than walk, was a lazy specimen with little affection for his horses.

grain. And, surprisingly, a large portion of the world's population still harvest their grain in just this ancient way.

One good thing about a lot of the work of a farmer—you get plenty of time to think. As a result, farmers have always been innovators and inventors, have always been coming up with ways of doing the routine chores a little faster or better or easier. So people have been concocting devices to make the harvesting of grain and other commodities more practical.

One of these, the vallus, appeared about 1,900 years ago. The vallus was a comb-like

16

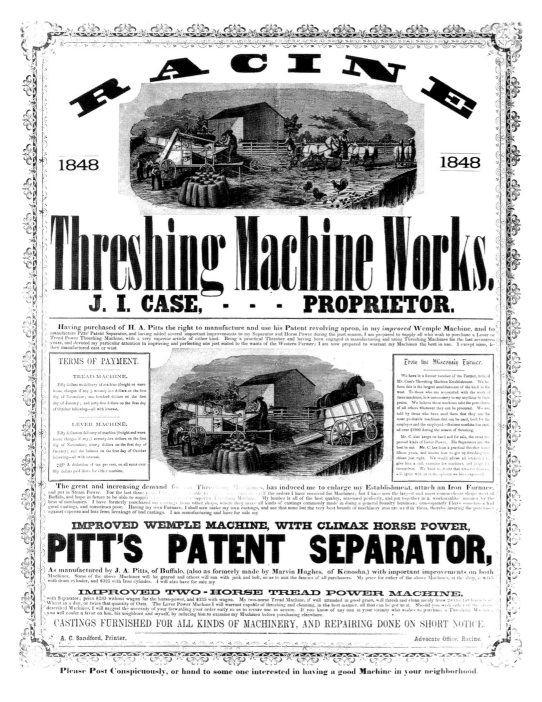

Jerome Increase Case knew the value of marketing and promotion; he used this handbill to inform farmers in what was then the "Wild West" (Wisconsin, Illinois, and vicinity) about his product line and business practices. The result was more demand than he could manage to supply, and the J.I. Case company quickly developed into a major force in American agriculture.

device on wheels that was pushed into a field of grain; the comb stripped the heads from the straw. The vallus wasn't widely adopted, and grain continued to be harvested with the sickle and scythe until the beginning of the 19th century when all sorts of experimental machines were introduced.

The American Revolution was part of a much larger revolution that encompassed Europe and much of the British colonial world. Along with the social revolution came the sudden, dramatic, Industrial Revolution that changed the lives of millions of people. That revolution hasn't yet ended and won't conclude as long as people harvest their grain with knives and thrash it under the hooves of oxen.

Silas McCormick and His Mechanical Marvel

One of the marvelous, revolutionary things to happen around the time of Ameri-can independence was a new attitude toward innovation and experimentation; everybody was doing it, and you didn't need an engineering degree (or even know how to read) to try. Interest in mechanical threshing began about 1740 with the publication in England of a description of an ancient Roman device for stripping the heads of grain from the straw. Many farmers experimented with the design and some publicized and patented their efforts.

The first American patent for a mechanical reaping system was granted in 1803, and another patent for a reciprocal cutter bar was awarded in 1831. Other elements of the mechanical reaper were being developed and perfected at exactly the same time Silas McCormick was busy with his version. By 1830 many people had tinkered with mechanical reapers and threshers, and lots of the contraptions actually worked—more or less. A decade before McCormick had the bugs worked out of his reaper a New York boy named Harvey May had developed a workable reaper but failed to patent or promote the design.

Silas McCormick usually gets the credit for the first mechanical reaper, but he had plenty of competition. He was part of an age and a process, and he didn't work in a vacuum. There wasn't a single, first, "grandmother" machine or inventor. In Britain several men developed and used mechanical reapers that did about the same thing as McCormick's; one of these was a minister named Patrick Bell who clearly had a functional machine at least ten years before McCormick, along with many others.

One of these inventors was a man named Obed Hussey. His reaper and McCormick's were quite similar—but Hussey's was on the market several years earlier. A lively competition in the press between Hussey and McCormick helped educate American farmers and get them thinking about this new technology. Verbal sparring matches, first in the *Southern Planter*, later in *The Prairie Farmer*, fascinated American farmers for years. Contests were common and not very conclusive.

What McCormick actually did was to assemble a horse-pulled machine with a

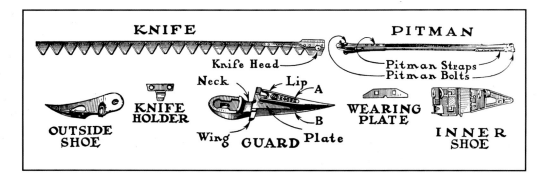

sickle-bar cutter and a reel, both powered by a wheel in contact with the ground—and he promoted his product and protected it with a patent. His real accomplishment wasn't the invention itself but his ability to get it into the field by the thousands. McCormick really invented the age of mechanized harvesting by writing letters to farm newspapers, sponsoring competitions, demonstrating his product, marketing and merchandising his reaper like a modern Madison Avenue advertising account executive. He sold on credit to allow farmers to pay for the machines from increased harvest revenues, he offered performance guarantees, and he publicized the machine aggressively.

It took about ten years for this new technology to catch on, but by about 1840 it was possible, practical, and popular for two men to cut as much wheat and barley in a day as five or six men did with the scythe or the cradle. The first part of the harvest was now mechanized.

Other people had been tinkering with mechanical ways to thresh the grain, including some designs that were used in the 1700s. One of these was a hand-cranked machine called a "ground hog," a spiked cylinder that broke up the heads. Other designs came along about the same time, some based on the flail idea, others that shattered the head with other techniques.

Within a very few years the whole technology of agriculture in North America and Europe was revolutionized. Mechanical reapers based on McCormick's pattern were refined and adapted. A mechanism to tie the shock of grain was added to the reaper and the binder was born. Fanning mills were added to threshers to make effective separators.

After the Civil War, American agriculture converted from small, subsistence operations to much larger and much less diversified farms. Instead of twenty or forty acres of wheat, along with dairy cattle, a small orchard, some livestock, and poultry, a farmer who moved west from New England to, say, Missouri would put all his effort into 160 acres of wheat, with very little left for substance needs.

Mechanical threshing arrived well before mechanical reaping. A canny Scot named Michael Menzies used a water wheel to power a set of flails in the mid-1700s, and another Scot developed the first rotary cylinder shortly thereafter. About the same time,

also in Scotland, Andre Meikle developed the concept even further with a machine that included a rotary cylinder and a mechanism that we now call "straw walkers," a device to move the straw out of the thresher.

These mechanical threshers made a big impression on the farming community of the British isles; farm laborers rose up in protest against the machines that were depriving them of gainful employ-

The John Deere No. 4 mower is ground powered—the friction of the wheel's contact with the soil turns the mechanism, and the rotary motion is converted to reciprocal linear action by these sturdy components.

ment. Owners of such machines were advised to destroy the machines or face the prospect of violence. Most of the machines were destroyed.

Then, in 1783, the British Royal Society of Arts, Manufacturers and Commerce published a translation of a description of an ancient Roman account of a grain head-stripper written about 100 AD. There were several attempts to recreate the harvester, mostly unsuccessful, but they inspired more development. In the great tradition of agricultural innovation, farmers pestered blacksmiths and tried out their weird contraptions in the middle of the night to prevent embarrassment.

Patrick Bell came up with the horizontal cutterbar, the reel, and the canvas apron in 1827—the first real reaper. It worked well, and in 1832 Bell's machine was cutting about twelve acres a day for just about four shillings; that was approximately four or five times what a man with a scythe could cut, at about one fourth the cost. But it didn't catch on, and Bell neglected to patent the device.

Silas McCormick experimented with his reaper for several years before obtaining a patent. He didn't really promote the machine until much later, but seven were sold in 1842, twenty-nine in 1843, and fifty in 1844. The success of his design was partly mechanical and part good promotion—and excellent timing.

Few were sold in the western frontier states of Indiana, Illinois, and Michigan until about 1845 when a McCormick representative appeared at the Chicago office of *The Prairie Farmer* to promote the reaper. It would, he said, cut an acre and a half every hour, and save a bushel of grain from every acre that was normally lost with the cradle. As described in *The Prairie Farmer*, the McCormick reaper required five men to operate: one to drive, one to rake, five to tie up the gavels. That meant that seven men could harvest about fourteen acres a day.

Next page, Big Green built their stuff for eternity—cast iron and lots of it. These two John Deere No. 4 mowers are no museum pieces, though, but working machines that are still—as was intended—horse powered. They are owned by an Amish farmer near Jamesport, Missouri.

19

They're cutting John Blank's 1929 wheat crop on a hot August afternoon outside Rosalia, Washington. The Case combine has its own engine, making the draft a bit easier for the horses, but it still takes twenty one of them to maneuver the Model H up and down these rolling Washington wheat fields.

The paper compared that to cradle reaping: seven men cutting, another seven men raking and binding, plus three men collecting the shocks would cut a total of twenty acres a day. The bottom-line comparison is that the reaper crew harvested two acres per man while the cradle crew harvested little more than half as much.

Of a competing design, one designed by George Easterly of Heart Prairie, Wisconsin, a farmer wrote in to *The Prairie Farmer* to report that the contraption "can cut a field all to destructive smash . . . it walks over the ground like an elephant."

Another hopeful inventor, Farmer J. Haines, showed up on the street outside the paper's office one day in 1849—with his reaper. He drove it up and down Lake Street, cutting an imaginary wheat field.

The Prairie Farmer described the virtues and vices of many of these machines, but the paper was critical of the habit of inventors to sell "patent rights" that allowed farmers to build their own harvesters rather than purchase the machines outright. Editor Wright said, in a September 1846 issue, "The farmer is not generally a mechanic, and where he has been at a great expense for a machine that will not go, he is not in a condition to remedy the evil. And though in cases he is ready to curse the knavery of the man who traded him his patent, he ought rather to curse his own greenness. If the farmer will resolve to see his implement work before he buys it, and then buy the

machine instead of the right to build it, he will save himself some wasted temper, as well as coppers."

Ambrose Wright didn't think too much of the McCormick model, and said of the raker on the machine that "he was a short, rose-faced, muscular man who rode backwards astride a sort of rail, with a stout piece of board attached to keep him from falling off. His legs were so short that he could only touch one toe; and he hung thus in the air, doing his work much as a man lifts a basket over a fence across the top of which he lies balancing himself. If it is not hard work for him, it was hard work to look at him."

Acceptance of any of these machines was minimal until the discovery of gold in California was reported. *The Prairie Farmer* predicted in July of 1849 that farmers would soon find that the labor force avail-

able to help with farm labor would be even worse than during the then-recent war with Mexico.

The bugs were pretty well worked out of the McCormick reaper by the time the American civil war began in 1861. The war took millions of men off the farm, but they were still hungry. The labor force dried up just as prices for commodities improved.

McCormick added a self-raker to the machine in 1862, allowing one man to accomplish the work of eight or ten harvesters in 1842. This design was an ingenious combination of reel and rake that first combed the crop into the cutter bar, then swept it off the platform into the neat bundles called gavels. The gavels still had to be collected by a "bandster" but even that backbreaking chore was about to disappear. It was complicated, costly at $175, but it changed the way farmers thought about farming.

Binder

By the 1870s the bandster had a seat on the reaper and tied up the gavels from his perch on the machine. This chore was mech-

anized too, first with wire ties, then in 1875 with twine knotting devices; another job disappeared and farming changed again. The twine binder remained an important farm implement right up to World War II and is still in use, still pulled by horses on Amish farms.

The binder was one of the first important labor-saving devices. It eliminated one of the most uncomfortable jobs of the harvest, and it helped the small, family farm harvest enough grain to feed animals, people, and to send to market. Particularly in the east and the damp portions of the Midwest, weather conditions allow a farmer a small window of opportunity to get the crop in before it shatters or spoils. With a binder, a farm family could successfully harvest about twice as much grain as the reaper in the same amount of time. Today they seem impossibly ancient and quaint, but they were important enough to stay on the market, almost unchanged, right up until 1947.

Mower

Another machine that was nearly as important was the mower, which was a device

Tom Coles has his horses well in hand as they circuit the field. A driver and his team working the old fashioned way is a beautiful sight.

that merely cut the crop without gathering it together in gavels. Mowers virtually identical to those of 150 years ago are still in frequent use by Amish farmers, carefully preserved and maintained to cut alfalfa, clover, and hay.

Stationary Thresher & Separator

Long before mechanical reaping was practical, mechanical threshing and separating were proven technologies in England and Scotland, with several designs on the market by 1810. These threshers were so popular and efficient that they provoked civil unrest among the farm laborers of Britain in 1830; about 400 threshers were destroyed during two years of unrest.

Next page, here's a typical sight in Holmes County, Ohio: A New Order Amish farmer harvesting oats with a three-horse hitch of Belgians. The Belgians are pulling a McCormick-Deering binder with a 10ft cutter.

Left, as the farmer circuits the field, his son and grandsons pile the bundles, or sheaves, into shocks. When grain is cut with a binder instead of a combine, it is cut a little sooner, when the moisture content of the grain is higher and the heads are less likely to shatter. The shocks will be left in the field for a week or so to dry out further and this will make threshing easier later on. Part of the art of the grain farmer is knowing exactly the right time to harvest.

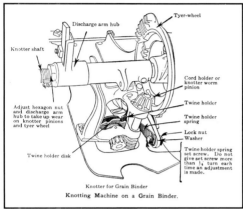

Knotting Machine on a Grain Binder.

These early threshers included all the basic elements of the threshing components of modern combines: a cylinder and beater section to flail the grain, shattering the heads and knocking loose the grain kernels, a set of screens to filter the kernels from the straw, a mechanism (called "straw walkers") to move the straw and chaff out of the machine, a system to agitate the straw, dislodging any retained grain, and one or more blowers to clean the dust, chaff, and straw fragments from the grain.

One of these machines was imported from England to the United States in 1788 and had a reported capacity of seventy bushels per day. The first were hand cranked, then horsepower was adapted to the chore. One of the first, designed by Jacob Pope, could thresh 150 to 200 bushels of grain an hour, attended by three men and a boy.

Threshers were in widespread use in the eastern and southern United States by about 1840 but their acceptance wasn't universal and midwestern farmers were initially skeptical of them. One of the factors that converted the skeptics was publisher John Wright and his weekly Chicago newspaper, *The Prairie Farmer*.

California and the Pull Combine

The first American combine of any practicality came from Michigan and the fer-

Ardenwood Historic Farm's Dave Cook supervises the Case thresher from his perch on top where he can monitor the progress of the wheat through the machine.

tile mind of Hiram Moore in 1838. Moore's combine worked fairly well when crop conditions were right, but that wasn't too often in Michigan where harvest time could often be damp.

But out here in the west grain growing conditions were virtually ideal—and grain harvesting manpower was virtually absent. Consequently any scheme or device, no matter how harebrained, was tested. One of Hiram Moore's Michigan combines was shipped out here on a clipper ship, around Cape Horn, and was set up very near here at Mission San Jose. The combine custom-harvested about 600 acres of wheat for three local farmers. The combines worked well in

The volume of straw produced by stationary threshing is one of the major disadvantages of the system; where do you put it all?

the dry climate, but the farmers failed to pay for the work; the experiment was a practical failure but a technological success.

The combine was used again two years later, in 1856, but an inexperienced crew failed to lubricate a bearing and the machine caught fire and burned.

But the need for a combine was still apparent and one of the men involved in the work with Moore's combine, John Horner, decided to build his own. His design put the

team behind the harvester, pushing the machine into the standing grain, but was mechanically otherwise a copy of Moore's design. Three were built, at great expense, and used in the fields east of San Francisco Bay during the 1860s. These combines could cut, separate, and clean about 16 acres of grain per day with just a three-man crew. And only one of them caught fire.

Within just a few years huge combines pulled by dozens of mules or horses brought

in the grain harvest across California, Oregon, and Washington. The big "bonanza" farms common here and in the Dakotas were ideally suited to this kind of machine, crop, and climate combination. The early ones were good for about fifty acres a day, and at

291045-C

Five men, twenty-one horses, one Case Model H combine on Frank Dyke's place outside Thornton, Washington, August 19, 1929. The grain is fat and shoulder high on the horses, and looks like it ought to be yielding about fifty bushels to the acre, maybe more—hard work for all concerned. The driver is using a "jerk line" to control just one of the leaders; these fine horses are some of the special breed developed in this part of the country, a cross between big draft horses and hardy little Indian ponies. This was the absolute high point of the glorious horse-drawn combine era, and within a year or two this kind of combine would be towed behind a noisy, smelly, cast iron tractor. *Take your picture, mister,* they seem to be saying, *we've got work to do. J.I. Case*

that rate many machines were required to cut the grain on a farm like Dr. Glenn's in Colusa County, a spread of over 60,000 acres with much of it in wheat, barley, and oats.

A popular alternative at the time involved a reaper that clipped just the head of the grain stalk with a minimal amount of straw. This harvester was called the "header"

THE FEARLESS.

Above, the invention of the blower system, called the "wind stacker," to direct the waste straw from the thresher relieved some of the burden from one of the toughest jobs on a threshing crew, but the dust and chaff still are a major occupational hazard.

and deposited the loose grain into big header wagons. When loaded, the wagons were driven to a stationary thresher and the grain pitched in to be cleaned.

The early combines were "ground powered"; the action of the machine was powered by a large wheel in contact with the ground. That meant the team not only had to provide power for moving the machine through the grain, but also had to provide the additional power for the cutter, the fans, and movement of the apron, the screens, and all the other powered components in the system.

When steam tractors began displacing the teams in the 1880s, some combine designers used the tractor to provide power directly to the combine mechanism, independent of a ground powered "bull wheel." These combines were huge; they were designed for, and only economically justifiable on extremely large acreages.

The first self-propelled combine was a monster built in 1886 by George Berry of Visalia, California, around a large Mitchell-Fischer 26hp steam tractor. With 4,000 acres under cultivation, Berry was motivated to find ways to bring in his grain crop more efficiently than the binder-thresher method, and his Model 1 SP combine was the result of a five-year development process. The machine used straw for fuel (free and readily available) and a flexible steam line delivered power to an engine on the combine assembly. Berry's combine originally had a 22ft cut, but for its third season had been converted to a 40ft header; it was the first machine to harvest more than 100 acres a day.

In a report to the *Pacific Rural Press* that appeared in the August 2, 1887, issue, Jacob Price reported, "When ready to start work,

Holt Model 32 combine detail: sack sewer's station, complete with skein of twine, ready for stitching up the sacks.

Mr. Berry asked me to take a seat by him on the tool box, which I did. Over our heads was stretched a large canvas awning some 20ft square that protected all under it. At the word of the engineer, Mr. Berry's brother turned on the steam, and away we went, cut-ting a swath 22ft wide just as easy as wink-ing. Just ahead sat a lazy fellow on a com-fortable spring seat, twisting a wheel like a car brake, right or left, guiding the ponder-ous machine to an inch. A fireman leisurely poked straw into the furnace to keep up steam. His hat hung nearby as if in a room, while his dog slept at his feet. In front of me was an engineer doing nothing at all, but alert and ready if required. On the other side of the boiler was a dusty looking fellow with an oil can in hand, squirting in a little here and a little there, looking mighty wise and pretending to work.

"On a separate platform on the outside of the separator was a sack sewer comfort-ably seated in the shade sewing sacks and dumping them off on the ground at inter-vals. Altogether there were six or seven men who looked to me like they had a soft thing, even though they were putting in sacks ready for market 30 or 40 acres of wheat a day."

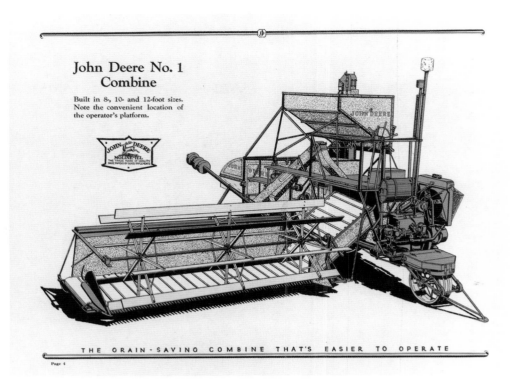

John Deere No. 1 Combine

Built in 8-, 10- and 12-foot sizes. Note the convenient location of the operator's platform.

THE GRAIN-SAVING COMBINE THAT'S EASIER TO OPERATE

Page 4

Above, the John Deere No. 1 combine appeared in 1929, with an 8ft, 10ft or 12ft header. *John Deere*

Left, Dale Palmer demonstrates the art and science of sack sewing. Properly done, the sack will weigh about 150lbs, be rigidly full, and securely fastened by the sewer. A firm sack is much easier to handle than a floppy one, and the "ears" tied off by the sack sewer provide perfect handles. Nine to twelve stitches and two tight knots and you're done—and on to the next one. Typical output for a sack sewer was about 700 sacks a day, about one every fifty seconds.

arator is of sufficient capacity that it can handle all the grain that can be got to it. I have averaged this season about ninety-two acres per day. I cut in two days 230 acres. It does not take any more men than I used last year to handle it, and it does about twice the work."

The steam era in agriculture was fairly short, but theatrical, and both the legend of the time and the machines have been preserved. Some of the steamers still earn their keep, too, but mostly they are relegated to the show ring. But steam tractors were the second major element in the conversion of Canadian and American agriculture from horse and hand power to the infernal combustion engine.

Then, in 1904, a gasoline engine was added for the first time, greatly reducing the need for big teams of mules or horses, but ground powered combines would be on the market for another fifty years.

Berry himself wrote about the first SP the following year after he'd converted the header from 22ft to 40ft:

"The machine I built . . . is as near perfection as a machine for the purpose can be made. The ground wheels are four feet face and six feet diameter. My header cuts a swath 40ft wide; it is made in two sections, and so arranged that it is handled with ease on very rough 'hog wallow' lands, at the same time making a clean cut where ever it goes. The sep-

Freshly filled sacks, each tant and firm, sewn by a practiced hand.

33

Holt Model 32 combine detail: reel drive mechanism.

Above, J.I. Case

Opposite, the design of the header section of this old combine uses the same canvases on virtually all grain harvesters from 1875 to recent times and the auger.

Chapter 2

How Combines Work

Combines are interesting machines. Essentially, there is very little difference between the machines of today and those of fifty years ago—in the fundamentals of how the grain or crop gets processed. And, if you look closely at combine components, there is a surprising similarity with the machines of 100 and even 150 years ago, in the way the components function.

Let's take a trip through a field of ripe barley aboard a typical combine of 1930—a John Deere, International Harvester, Case, Baldwin, Allis-Chalmers, or any of the others that all worked on the same basic principals.

Our rig will be pulled through the field by our sturdy tractor, the same faithful beast that we use for plowing. We only need three people to bring in the crop: someone to drive the tractor, someone to operate the combine, and somebody else to bring a wagon out to the field to collect the grain from the grain tank on the machine. Both the wagon and the tractor can be driven by members of the farm family—a boy or girl of ten or older often got either chore, although the farmer's wife, hired man, or neighbor might do the job as well.

Opposite, a detail of the reel drive mechanism on an International Harvester binder.

Well before harvest time, about when the grain reaches the "milk" stage, we will go out to the barn to prepare the machine. All the many bearings and oil cups need to be filled, all the components need to be inspected. We will probably pull the machine out, hitch it to the tractor, and give it a dry run. The sickle bar will get sharpened and oiled, some sickle bar teeth and cylinder and concave teeth may need to be replaced. We will pay particular attention to the drive belts and chains; they will probably need adjustment or replacement.

As every grain farmer knows, the crop ripens through five progressive stages during about twelve days. The first is called the "milk" stage, about sixteen days after the heads form; that is when the kernels are full of the milky sap of a rapidly growing plant; it is possible to harvest the crop now, but it will only yield about 22 bushels to the acre. Within two days or so, however, the kernels will begin to firm and mature; this is called the "early dough" stage, about eighteen days after the plants head, and the kernels are still rapidly developing; you get about 26 bushels to the acre if you have to harvest now. Just three days more and the grain should show signs of the "dough" stage, a much larger and firmer kernel; if you harvest with a scythe or sickle now you won't loose much to shattering, but you will only get about 32 bushels to

When Holt Brothers decided to rebuild their Model 32, they went whole hog and brought it back to working condition, including the engine. It started easily, and with a few minor adjustments to the carb, ran merrily.

29258

CASE
COMBINE
Hillside Type

Left, during the summer of 1929, Thomas Morris harvested 1,700 acres of South Dakota barley with this Case Model P pull-type combine out at the edge of the Great Plains. *J.I. Case*

Above, brochure cover, 1920s. *J.I. Case*

the acre. Twenty-five days after the grain forms a full head it reaches the "stiff dough" stage and will produce about 33 bushels to the acre; the heads become somewhat fragile and begin to shatter. Finally, twenty-eight days after forming heads, the grain will be fully ripe, yielding about 35 bushels to the

acre, but it will be quite fragile and likely to shatter—perfect for combining but too late for reaping without substantial handling loss.

We want to cut our wheat, barley, or oats on a clear, dry day, and we will wait to begin until the sun is well up and the dampness of the night has been burned off. Then we all take our places, and the tractor driver maneuvers this grand assemblage to the cor-

Right, cutaway view, Case hillside combine. *J.I. Case*

Right center, cylinder and threshing section cutaway diagram. *J.I. Case*

ner of the field. Once we are properly lined up, off we go, opening the field in a clockwise direction. The tractor needs to maintain about three miles an hour for the combine to work properly, and the soil needs to be dry for the "bull" wheel to have good traction.

Heading

As we approach the edge of the field the whole machine comes alive, quivering as if with anticipation. The sickle bar knife oscillates, glittering with fresh oil and just-sharpened edge, clean in the morning light—then it takes the first bite of the standing grain. Above the knife rotates the reel, sweeping the heads and straw back, into position, guiding the cut stalks back to fall on the platform. The grain marches, up, into the combine, and disappears into the machine.

From your perch on top of the combine you can easily raise or lower the header section to accommodate the gentle swells and troughs of the land. The idea is to get every last head of grain, missing none. The tractor driver and the combine operator have to function as a team, keeping the combine fed without overloading it. Weeds can plug it up and it is your job as operator to be sensitive to any disturbance in the machine that might mean a breakage or other problem.

Below, separation section cutaway diagram. *J.I. Case*

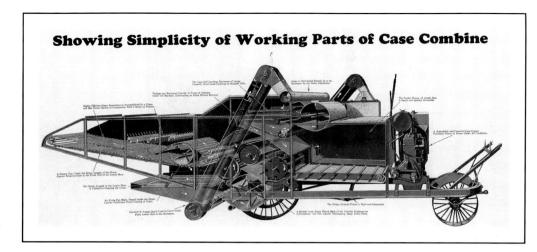

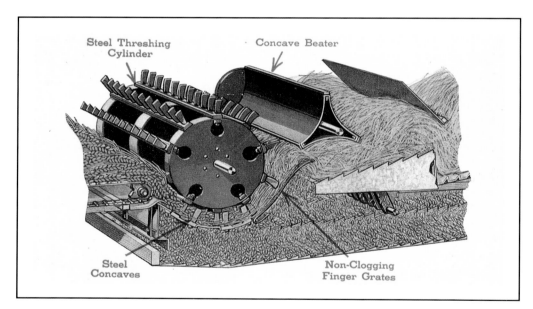

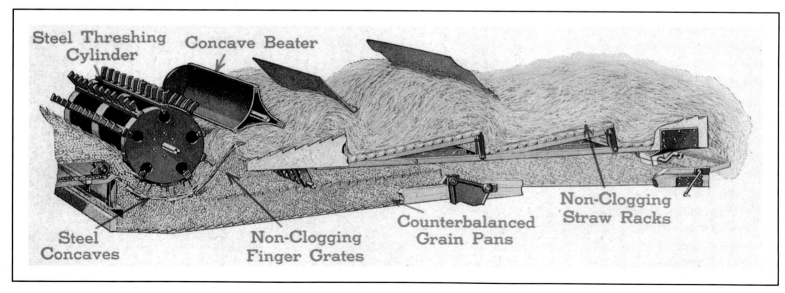

the straw, pulling the kernels out and away, and pushing the loose crop back into the machine for further processing.

Separation

The process of removing the kernels of grain from the straw is called "separation," and most of it happens right at the cylinder. About 90 to 95 percent of the grain is knocked out of the straw at the cylinder, cascading down on a large sheet of metal called the "grain pan." But that other 5 to 10 percent of the crop might be your profit margin for the year and you don't want to let it get away, so threshers and combines are designed to get it all.

The straw is pushed, shaken, and fluffed in a jerky rotary/horizontal/vertical shimmy, back toward the exit at the back of the combine, allowing the remaining grain to trickle out of the straw, down through the straw racks, onto the grain pan that runs the full length of the machine. The effect is very similar to the ancient technique of tossing the straw into the air and striking it with a pitch fork. An access panel at the top of the machine lets you peek inside to insure that the process is working as advertised.

Cleaning

The grain is now separated from most straw and chaff, but is still mixed with dust, dirt, and fragments of straw. It is cleaned in two parts of the machine, the lower and upper cleaning "shoes." In both, large fans blow air through the grain kernels as they are allowed to fall through sieves and screens; the lighter straw and chaff are blown away, the heavier kernels retained. This process is repeated twice, and you, as operator, can observe the final cleaning process through an access door at the top of the combine. Finally, the clean grain is transmitted by an auger to bags, to a grain tank, or to a wagon alongside. The straw is spread behind the machine.

Harvest time with a combine comes later than it does for our neighbors who are still using a binder. They will cut about a week before we will, to reduce the loss of grain from shattered heads. Their kernels will be just a little smaller and less mature as a result, and we will get more bushels from an acre. That's one of the reasons we spent that $1,500 for the machine. Another reason is that our neighbor with the binder and the thresher will spend three weeks bringing in

Threshing

The grain feeds head first into the threshing cylinder, out of sight, the second phase of the process. The cylinder looks very much like the cylinder of the ancient groundhog machine of the 1700s, but with steel replacing wood for the cylinder and teeth. The teeth rotate through another set of stationary teeth, the "concaves." The tip speed of these cylinder teeth are about 5,000 feet per minute, fast enough to strike each head several times during its passage through the threshing section, but not so fast that the kernels are broken in the process. Much of the grain falls out of the heads immediately, along with plenty of dust, dirt, fragments of straw, and the "beards" and fragments from the heads of grain. The action of this cylinder is a bit like a comb or rake that rapidly sweeps through

his crop, while we only need five days. And he's got to pay and feed a crew of eight or nine people—unless he's got a lot of kids. The combine costs more to start but less to keep over the years.

How to Run a Binder

How does a binder work? Simple—by magic. It's a complicated machine that appears to have been designed and built by angels. You can study it, watch it work, see it cut, gather, bind, and dispense neatly tied bundles of grain down the length of a field and you will probably reach the same conclusion a lot of folks do: it does the impossible.

Well, of course it isn't quite impossible but it did take a long time for people to invent a practical knotting mechanism and adapt it to the reaper. Several men were responsible, all coming up with key patents

Cleaning section cutaway diagram. J.I. Case

during the 1860s and '70s that collectively solved the problem. The most important of these is probably the 1864 patent of a man named Jacob Behel from Rockford, Illinois, who invented what is today called the "bill-hook," an L-shaped component with a hinged gripper that looks and works very much like a bird's beak.

The binder mechanism collects and compresses the sheaf of grain, then—in less than a second—wraps the sheaf with twine, ties the knot, and kicks the sheaf out. As long as the contraption is working, you don't have to worry about it—there are other things to worry about!

First, make sure your grain crop is ready for the binder. The traditional way to judge is to study the uppermost leaf on the straw; it ought to be just turning a golden yellow. You want to avoid too much green (poor development) or too brown (prone to shatter). Walk out into the field—you should be out there often in the weeks before harvest—and

pinch off a few heads and thresh them in your hand. Your palm serves as the concave and if you blow on the chaff, your breath will be the thresher fan. When those heads start to get a little brittle snap as you rub the kernels out, it's time to bring in your crop.

You ought to have already been out to grease the machine, particularly the pitman arm—a critical part since it provides the motion for the sickle bar knife. Work your way around the machine with a wrench, checking and tightening any loose bolts. Be absolutely certain that there is enough lubrication on the machine. You ought to put a fresh roll of twine in the canister, even if there is plenty from last auger; mice like to set up housekeeping inside the balls, and the moisture from winter and spring rains may have induced some rot. Thread the twine, then pull the machine through a cycle by hand, just to make sure the knotting and tying mechanism is functioning properly. You can't really have too much twine

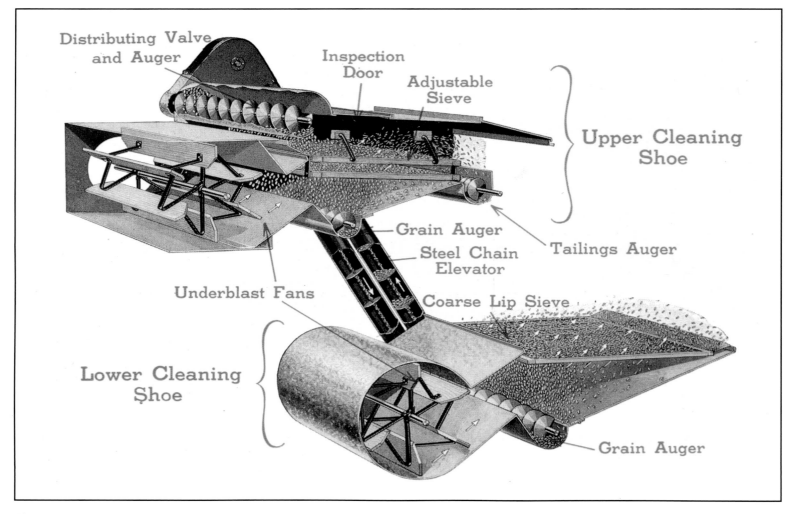

aboard, and there is room for two rolls on the machine, so make sure you have a spare. Each roll will bind the grain from about one acre of crop, depending on its condition.

The binder hitch keeps the weight of the machine entirely on the ground; the horses just pull straight ahead. Even so, the animal's up and down motion will affect the cutter bar a bit, adding a little up and down unevenness to the swath.

Bring the horses up, hitch them up, and walk them out to the field, allowing them to get used to the rattles and rhythms of the machine. On the first day of binding they will probably be spooky and uneasy about this thing they are towing—and their natural reaction to something that scares them is to run away from it. This is not a happy experience for horse or rider when it happens with a binder.

You perch on a typical farm implement seat, about 20ft behind the horses, with four levers in reach and with two reins in hand. One lever adjusts the height of the frame relative to the ground, another adjusts the "grain" or outrigger wheel to keep the cut level. You want to cut the crop so you get enough straw to form a good bundle, long enough to compact but not so long that you and the threshing machine have to struggle with unnecessary extra straw. You don't want the straw too short, either, since that would allow the heads to feed into the elevator section at all angles, preventing the binder from forming neat sheaves at all. Another lever controls the height of the reel—try to have it hit about 6in below the head, and be sure that the reel isn't too far forward relative to the cutter, either, or the grain will get

knocked down before it gets to the knife. Finally, a fourth lever controls the angle of the machine relative to the ground—adjust this one so the binder tilts forward a bit to help the grain lay down flat on the apron. Now make sure the twine tensioning device is adjusted correctly and the canvases are all tight.

As you start cutting the crop you must monitor all the systems at once, somewhat like a fighter aircraft pilot. Are the canvases all moving properly? Are you cutting too high or low? Is the speed of the horses too fast or slow? Is the twine paying into the knot mechanism properly? Are the sheaves being bound correctly? Is the machine tying good knots?

Top view, Case combine, 1920s. *J.I. Case*

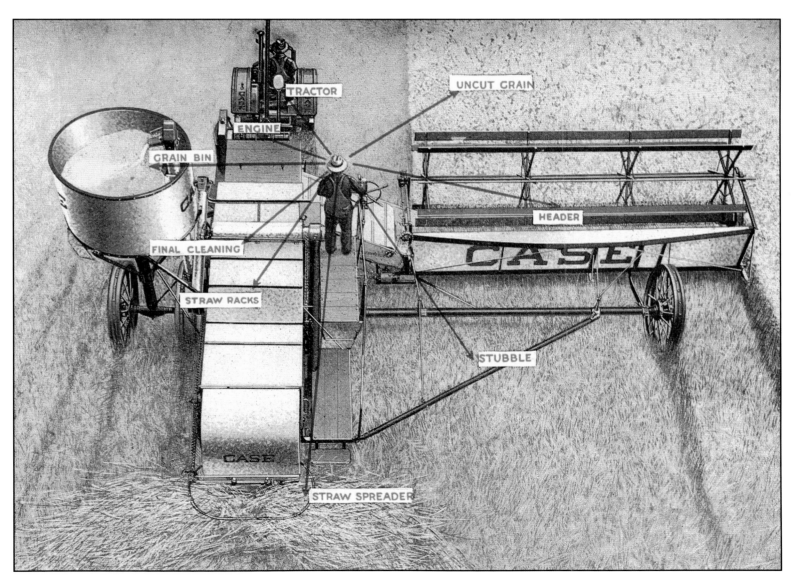

43

(1) Cutting—The Header cuts high or low, follows the slope of the ground and gets all of the grain (see illustration on page 4).

(2) Threshing—Kernels and seed are swiftly and throughly loosened from the heads by the all steel threshing cylinder and concaves.

(3) Separ throug grates out by

CASE Combine
(*irie Type*)

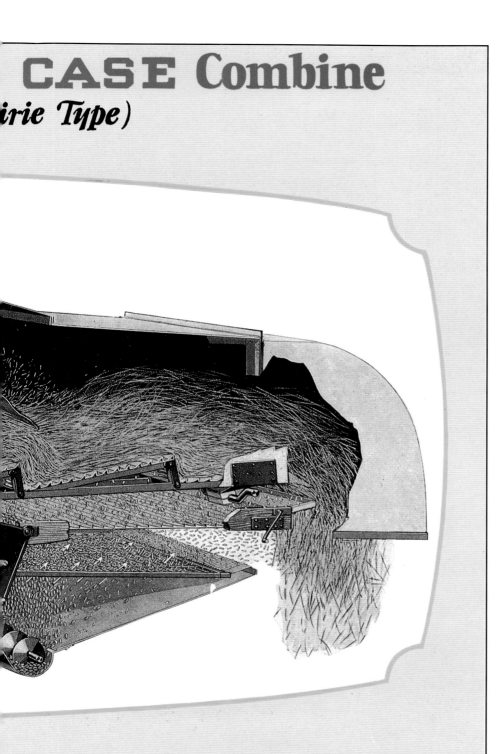

Grain falls out
icaves and finger
ompletely shaken
ating straw racks.

 Cleaning—Impurities are readily removed by two cleaning shoes, with the final cleaning under the eyes of the operator.

Tom Coles and the team are just getting to know each other, and neither is quite sure who's in charge. Tom is pondering his next move, and the horses are doing the same.

International Harvester binder.

Left, Tom Coles aboard the International Harvester binder during Bygone Farming Days near Sacramento, California.

Above, International Harvester binder.

Right reel detail, International Harvester grain binder. The reel position is adjustable up/down and fore/aft. Ideally it will smoothly strike the straw about 6in below the head, guiding each shaft back into the knife, then sweeping the crop back onto the platform, to be carried to the aprons.

Twine problems are the principal frustration of a binder operator, and this one is currently thinking black thoughts and wondering if maybe one of them newfangled combine machines would make sense next season.

Four levers on the binder adjust height, tilt, cutter bar height, and reel position.

IT CUTS THE THRESHING TIME

"After years of worrying and tinkering with various kinds of belting, I can appreciate this new Goodyear Klingtite Farm and Tractor Belt of mine. It's going on its second year now, and hasn't developed a flaw. It cuts my threshing time and saves my men. I am looking forward to a lot of good service out of my Klingtite Belt."—ANTON HAGEN, Farmington, Minnesota.

You, too, will get your threshing done a whole lot easier, faster and at lower cost with a Goodyear Klingtite Farm and Tractor Belt. This belt is scientifically designed and built for farm power duty. It holds the pulleys in a powerful, slipless grip. It runs loosely, favoring the engine bearings and making the most of fuel. It is weather-proof, moisture-proof, trouble-proof. Eliminates re-setting because it does not shrink and is subject to only the minimum of stretch. Outwears ordinary belts by a wide margin of efficient, economical service. Requires no dressing; needs no breaking-in.

Goodyear Klingtite Farm and Tractor Belts are made in endless type only for heavy duty. Other belts, in cut lengths, specially constructed for lighter drives. They are sold and serviced by all Goodyear Mechanical Goods Service Station Dealers and by many progressive hardware dealers the country over

Copyright 1927, by The Goodyear Tire & Rubber Co., Inc.

KLINGTITE BELTS

Combine Economics
 be with this.

Chapter 3

Combine Economics

How To Make The Farm More Profitable

When farmers discovered the advantages of the combine (and their conservative nature was satisfied that the advantages were genuine), the new machines were rapidly adopted. Here's why:

The combine saved a tremendous amount of time. Binding and threshing are two big, slow, labor-intensive jobs. A farmer with 200 acres in grain would need around six days to just cut and bind his crop; it would have to be allowed to dry, usually in the field, for about two weeks, then another four days would be needed to thresh it. We wait until the grain is fully ripe, then bring our 200 acres all in during just five or six days, about half the working hours.

The crew is much smaller and the work much easier. Three or four people can harvest about 40 acres a day. That works out to about 45 man-minutes per acre, while binding and threshing require about 3.6 hours per acre. The actual out-of-pocket cost, according to USDA, works out to be about half of the costs of binding and threshing. This dispensed with the ragtag army of itinerant threshermen who once spent the summer following the harvest—by all accounts, a very seedy lot. One of the virtues of this system, according to an ad for Case combines of the 1930s, was that with combining, "The combine eliminates cooking and board bills for such large crews! No more long weary

On the road again—but where are the rest of the horses? Well, the team isn't as big as was typical for this machine, but just about every draft horse in San Joaquin county was recruited for this event.

It may look quaint today but this is state-of-the-art technology within the memory of some of those present to watch this reenactment, and such machines once harvested 100 acres a day.

What the Ladies Say About the Combine

Gentlemen:

I am more than glad to state, from my point of view, the labor the Case combine has saved me. Before we purchased a Case combine there were long hours from 4 AM until 10 PM cooking and preparing for 12 to 15 men. It was necessary to store up an enormous amount of supplies for the harvest, but now it's the same throughout the year. Why shouldn't I praise the Case combine when it means hours of pleasure instead of labor?

Mrs. W. S. Lyons
Hoover, Texas

Gentlemen:

It is with pleasure I try to express to you in words my appreciation of the Case combine, It simply means that the dread is entirely dropped from harvesting.

Now we have but one extra man to help my husband and son harvest and thresh, where formerly we had around fourteen men. This means no extra beds to be made up and often take up before I could serve breakfast in the mornings.

It means I do not have to cook cabbage and potatoes by the peck. It means cool cooking, for I now use the oil stove entirely throughout harvest time. The old way, with a large crew, it was necessary to have more stove room, therefore the use of the hot range.

It means fewer dishes, pots, pans and kettles to put away. And, by the way, I now go right ahead with my wash days during harvesting.

We also have about an hour's more rest now, as we can retire earlier and do not have to get up so early. Last, but not least, it means only one "hitch" at the job. "When it's done, it's done"— no header crew, then later a thresher crew. I indeed feel fortunate in being a farmer's wife since a Case combine does the work.

Mrs. F. D. Mason
Texhoma, Oklahoma

Baking for extra help is eliminated when a combine is used. *J.I. Case*

Only the farm woman realizes the extent of the work involved in cooking for a threshing crew. *J.I. Case*

Four o'clock and the lady of the house is out of the kitchen. "No words can express my gratitude," says the wife of an owner in Hooker, Oklahoma. *J.I. Case*

Gentlemen:

I would not hesitate to give your company a few words of appreciation regarding the wonderful Case combine we purchased from you last season. Possibly the most benefit the housewife receives from one of these is that of elimination of the tramps and undesirable help one needed to put up with when cutting wheat the old way. Two years ago when we cut wheat with the header we hauled twenty-five men out from town.

Last year with the Case combine we employed two very respectable men who started and finished the job. We did not lose a moment's time on account of the Case combine. We cut nearly 500 acres, while our neighbors who have machines put out by two different companies cut only 250 acres—and then we finished two or three days sooner.

Below is a general summary of what I believe we saved in each day's run with our Case combine:

3 men—9 meals @ fifty cents each	$4.50 per day
3 men—wages @ $4 each	$12.00 per day
1 hired girl with meals	$3.00 per day
2 bushels wheat per acre/50 acres	$150.00 per day
6 cents per bushel on 500 Bu. per day for threshing	$30.00 per day
4 days straw scattering on 50 acres @ $1.50 per day	$6.00 per day
total	$205.50 per day

Mrs. Geo. Eatinger
Raymond, Kansas

hours of cooking for the these large crews of hungry men, and no strange laborers in the home. Many farm women, and their husbands too, dread the approach of harvest and thrashing because it means strange men about the place; a combine helps protect you and your family."

• Less wasted grain: Grain is inevitably lost when it is handled. The US Department of Agriculture (USDA) calculated that about seven percent of the crop was lost when binders and threshers were used, while about 4.5 percent was lost with the typical combine.

• Lower direct costs: USDA calculated that it cost a farmer $3.82 per acre in 1930 to bind, shock, and thresh each acre of grain, on average, while a combine's costs were about $1.72. In a comparison of costs with binding and threshing, combining saved about 80 percent of the labor involved to harvest, and out-of-pocket cash costs for operating the combine were only $ .39 per acre (fuel was twelve cents a gallon at the time). As the study discovered, the cost of twine for the binder cost more than gasoline for the combine! The binder/thresher hard cost was calculated at $1.79 per acre.

• Higher Yields: The University of Nebraska calculated the yield loss associated with early harvest and found that grain cut at the "dough" stage (about twenty-one days after it has reached full head) will produce about eight percent fewer bushels than at the fully ripe stage seven days later. That is a differential of about three bushels.

• Better price: Although a farmer with a binder cuts his grain earlier, it goes to market later. That made it easier to choose when to sell. Farmers found that combined grain tended to be of higher grade with less dockage—and that they made more money from each acre, on average.

Canadian and US farmers were often considering these factors in the years after World War I, as compact, efficient, affordable combines became available. As Case asked in a promotional brochure from the 1920s, "Will it Pay on Your Farm."

Will it Pay on Your Farm?

That's a question every grower of threshable grain and seed crops should ask and prove to himself. It has been estimated that farmers with no more than 50 to 100 acres of threshable crop can save money by owning a Case combine, especially if combining is done on the side.

Combining for Profit:
What the Fellers Think

Thresh at the best time for bigger profits. *J.I. Case*

We cut 1000 acres in 1928 and 1850 this year and the combine is in fine shape, ready for more grain. This year our best day's run was 90 acres, but we started early and ran late.
A. J. Foster, Consort, Alberta

I have a son fifteen years old who drove the truck and hauled all our wheat. I ran the tractor and we hired a man to run the combine. We sure would hate to go back to the old method of harvesting and have a crew of twelve or fifteen men around for a week. There are no long hours for the wife with only one extra man.
T. P. Forster, Venango, Nebraska

I figure I can cut and thresh with a combine at the same cost as cutting and shocking it in the old way.
Lee Reece, Mifflinvale, Pennsylvania

Once over with the combine and the job is done and you can get the grain to the elevator early. You can do more early fall plowing and can enrich the soil by plowing under all the straw.
Will R. Mehren, Mott, South Dakota

I sold my wheat from the combine for $1.29 per bushel and had no dockage. I also finished combining the same day my neighbors started threshing. The neighbors who threshed the old way received for their wheat from $1.10 per bushel down to 70c per bushel, due to damage in the shock from wet weather.
L. E. Rudisaile, Lamar, Missouri

The combine sure is the thing for quick harvesting and threshing. I had more custom work than I could do and have my combine all paid for. I got from $2.75 to $3 an acre for combining. Now is the time to own a combine when custom prices are still high, as they will pay for themselves.
Henry Holland, Glenham, South Dakota

The thresher owner is independent and labor is saved for both men and women. *J.I. Case*

Spearville, Kansas
November 7th, 1927
Gentlemen:

I cut my own wheat with the aid of my two small boys—one 14, the other 10. I run the tractor, the 10-year-old boy runs the combine and the 14-year-old boy hauled the grain to town with the truck, excepting what I put in the bin. I hired a man at $5 per day for scooping. When I cut for a neighbor, two men hauled with wagons and my boy hauled with the truck, and we kept them busy.

My wife was sick in bed with the flu. My 12-year-old daughter did the housekeeping and cooking alone, so you see we harvested and threshed all alone, which we could not have thought of the old way. Me for a combine!
Yours truly,
Ed Sanko,

Turon, Kansas
October 23, 1927
Gentlemen:

Went through harvest without any combine trouble and not a penny expense for repairs. My women folks remarked, "My! This doesn't seem like harvest. Everything runs along so smooth and by night we know just where we stand, how much wheat we have cut and threshed, and have the money in our pocket. Just have three men to cook for, and when you are through, you know it is over."
Yours truly,
F. O. Bunyan

Manchester, Oklahoma
November 11, 1927
Gentlemen:

My father, brother and I bought the combine together. We three did all the work ourselves. We had an elevator at the bin, so one man could handle the grain and the other two ran the tractor and combine. Our grain was put in the bin for less expense than our twine bill was the year before, figuring gasoline and oil.
Yours truly,
G. E. McDonald

This frozen moment in time is August 7, 1930, and the place is the Gunkel farm outside Hillsborough, North Dakota, and the economy of the US and Canada has just crashed. Commodity prices have gone to hell and the terrible thirties, with drought, dust, and disaster, have just begun. But the wheat is coming in anyway with the help of the latest in farm equipment—The Case Model L tractor, with PTO drive powering the binder. *J.I. Case*

A fair way to figure on combine profits is as follows:

Take first your binding and threshing costs, include the yearly costs of owning and operating the binder; include costs of twine, costs of shocking, etc. Add to the above your threshing costs—not only the money paid to the threshermen but costs of labor and equipment to haul the bundles to the thresher and the grain to the bin.

Now figure what combining would cost. Get the price of a Case combine from your dealer and divide it by 15 to 20 years' life to get your yearly first costs. Add to this your estimate for repairs, grease, oil, and fuel as well as the cost of operating your tractor during the combining period.

Deduct your combining costs from the binding and threshing costs to get total savings from combine ownership. Add to these savings your probable returns from custom combining at an average figure of $2.00 per acre. The result will be the approximate total savings and earnings from owning a Case combine.

The above results will be cash savings only. It will not include the many other advantages of Case combine ownership such as: *higher yields per acre, less hard work for the womenfolk, more time for other work, etc.*

The windstacker not only removes the straw swiftly but also delivers it in a neat pile. *J.I.Case*

Next page, 1927; this hillside combine's young header-puncher is setting the cutting height. He's the junior man on the crew, a boy of about fifteen—who would be in his eighties today, and may yet recall this glorious day. *J.I. Case*

Above, the header-puncher tries to cut as much of this downed oat crop as possible, but the grain is in bad shape in this late 1930s photograph. It was a bad decade for farms, farmers, and for implement manufacturers. *J.I. Case*

Left, harvest time, 1941. Pearl Harbor is only weeks away, and these men are still wondering if the US will be able to stay out of the war. When the war comes, this little combine, with its crew of three, will get a five-year workout.

This little Model A Case combine only needed a crew of two: the operator on the combine and a man to drive the tractor. *J.I. Case*

Chapter 4

Driving the Horses

The Work of the Month

Driving a team of horses is a little like driving a car . . . a car with a sense of humor, paranoia, adventure, and a mind of its own. You put a team in gear by taking the slack out of the reins, release the brake on the binder, then step on the accelerator gently by talking to your horses. What you say to them depends on how they (and you) have been trained; the Belgians at Ardenwood know "UP!"

You need to look well ahead of the horses, at least five to ten feet ahead of the horse next to the grain. That horse, the one next to the crop, is your key to efficient cutting. And since there is about fifteen feet between where that horse is and where the knife is, you have to make slow, gradual course corrections to avoid over steering. You want to drive as straight as possible, and the best way to do that is to look far ahead. Use the reins in a gentle, fluid movement to communicate with the horses. The animals will give you what you want as long as you can convey your intention to them.

Opposite, a three-horse hitch of Belgians, like these on an Amish farm in Ohio, provides a kind of partnership and companionship not possible with tractors. As more and more people are discovering this, and the other benefits of farming with horses, the demand for farm horses and horse machinery is increasing.

With every team you have a "worker" and a "lagger"; one horse will pull until it is just about ready to collapse—and the other horse is happy to let it take all the load. The horses know their names, so you can talk to them, and they will respond. After two or three trips around the field the horses understand what they are supposed to do and will, if they are a good team, function pretty much on autopilot.

How do you turn a corner? Stop the team right at the corner, just as the knife finishes the cut, then with just a light pull on both reins command, "come around!" and the horses know to step sideways, pivot the machine 90 degrees until you are lined up on the next side of the field, and you are ready to go again. Actually, you will talk to the horse on the outside first, since that one will have to start the move; with the Ardenwood Belgians, that's Charlotte, and you say, "Charlotte, come around . . . come around," then, "whoa!" as they bring the binder into position.

When one of the horses starts to lag you can spot the tug lines go a little slack and the double tree will pivot toward the "worker" horse, then you have to admonish that horse to tighten up and pull its share, "Charlotte, UP! UP!"

Bundles accumulate on the binder until you decide to drop them; a foot pedal operates the release. The pick-up crew's job is a little easier if they can collect half a dozen bundles

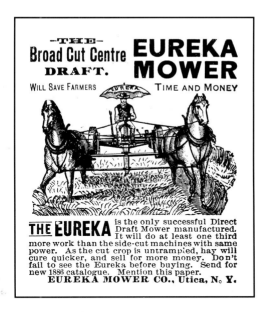

at a time, rather than each one individually.

Runaways and "bustups" were a common experience with teams. Horses by nature are ponderous, scary animals—their natural response to something unfamiliar or frightening is to run away from it, then to stop and look back at it. That's particularly true with horses with lots of energy and that haven't been worked recently. The more you work a team, the better they are, the better their behavior.

Combining with Teams

What was it like to combine grain back around the turn of the century? It was fun and very hard work.

The day would begin around 4 AM. While the older men—the family patriarchs and the hired threshermen—attended to maintenance chores and prepared the combine, one of the kids would saddle a horse and try to collect the mules from the pasture where they usually spent the night. This generally took a while because the mules knew how to play the game. But sooner or later the mule wrangler would herd the animals into the barn where the rest of the crew would be waiting.

Individual animals got sorted out for the specific teams: normally thirty-two would be used for the combine, another sixteen for the header boxes, plus several spares. Since the animals usually worked the same team each day, they soon learned the routine and went to their assigned stall without much encouragement. You clipped a halter chain to the animal's halter, then fed **it** breakfast.

Breakfast for a hard working mule is two or three quarts of grain—wheat, oats, or barley. While the animals ate, the men harnessed them. The harness for each animal will be on a peg by the stall, and since it is already adjusted for the individual mule it is a simple matter to put on. First the collar is slipped around the shoulders, then the "hames" are fastened at the top, then the rest of the harness is placed on the animal's back for easy access. The belly band and the crupper are fastened, then the bridle is slipped onto the mule's head. Your first mule is ready. It takes about half an hour to harness. Then you are allowed to put on the feedbag yourself.

Below, working with a team of horses can be a real pleasure, and it's sometimes a lot easier than with a tractor. A tractor can't learn what it's supposed to do but a horse can. After only a couple of times around the field, the team will learn exactly what is expected of them. After that, the driver only has to guide them and offer encouragement once in a while. Plus, the horses are much better company than a tractor.

Before hitching up, the team is watered. The mules are hitched eight abreast, then led to the water trough. When they're done, you back them away, then hitch them two abreast, then lead your eight animals out to the combine. After a few days the mules will know just where to stand when you connect the tugs and connect the halter chains to the center chain. Finally, the jerk line is run up to the near leader through the clips on the harness of the mules on the near side.

With the combine crew aboard and the grain dry enough to harvest, it is time to begin the day's work. The combine driver, known as the "mule skinner," puts the team in motion with one long, loud whistle; the

lead mules lean into their collars, take up the slack in the chain, and the rest of the team starts to pull. The mule skinner really only controls the leaders—and sometimes only one of the mules, the jerk line leader.

Mules are tough and smart, and sometimes quite independent. When they feel like cooperating, they can often make a much better draft animal than a horse. A horse, even a smart horse, is not very intelligent; given unlimited grain, a horse will eat itself sick. A horse will also work itself to death, literally dropping in the traces. A mule will do neither of these things. But a mule can have a sense of sport and adventure—and, even worse, a sense of humor; none of these at

Hitching the leaders, a pair of big, handsome, and dependable Belgians before a trip around the field.

tributes are the sort of thing you want in a team of thirty or forty animals hauling you and your combine across this year's wheat crop.

Mules are slower than horses, though, and that sometimes created problems, particularly on sloping hillside fields. Generally, a farmer with a quarter or half section (160 to 320 acres) tended to use horses while the bigger operations often used mules.

Here's an unusual sight in Amish country—Percherons instead of Belgians. The horses' names are Matt and Mike and are the delight of the young man on the hayrake, who is helping his uncle to make hay. The hay is too damp to bale, so it will lay in the field and cure for a week or so.

Horse Power

A combination of factors resulted in an explosion of agricultural products in the latter half of the 19th century. One was the successful development of mechanical harvesting equipment, another was the opening of vast tracts of farmland in the Midwest and West, and another was the development of efficient transportation systems. But the foundation of the new productivity was the draft horse.

This new kind of agriculture needed a new kind of powerplant. Until steam engines were adapted to mobile use, that meant that horsepower came from the traditional source, the horse.

For the little acreages of wheat and barley and corn on eastern farms, one or two horses were all a farmer needed for plowing and reaping. That same team of Shires or Belgians might power a thresher or a separator after the harvest on a treadmill or windlass. But once farmers started growing a whole quarter section of wheat out in Kansas or Minnesota it took more than a matched pair of draft horses to bring in the grain.

Farmers, particularly in the west, used larger and larger teams of horses. Some of these pulled the first combination reaper and thresher, a Michigan design that brought in the 1854 harvest on the rolling hillside fields around Mission San Jose.

By 1869 California was the eighth largest wheat producing state in the union, producing 69 billion (that's what it says) bushels on 2.5 million acres of farmland. One farmer owned a farm with an area of over 100 square miles, and much of it was in wheat. It took big crews and state-of-the-art harvesting equipment to bring in such quantities of grain, and a distribution and storage system to get it all to market.

Left, there are about 20,000 Belgians like these in America and Canada today, many still used on farms in the traditional way. Charley and Charlotte pull binders, plows, harrows, and planters on Ardenwood Historic Farm.

Above, draft horses, like the beautiful Percherons, Matt and Mike, are the pride and joy of farmers all around the country who know the pleasures of working with these magnificent creatures.

Above, this photo clearly shows the hay being turned over to dry in the sun. The rake that Matt and Mike are pulling is of an uncertain pedigree, but that's not surprising. Amish farmers frequently convert implements designed to be used with tractors for use with horses.

Raking hay on another Amish farm with a fine
pair of Belgians.

Above, Dwight Gilbert and his team of Percherons prepare to take the International Harvester binder around the field again. Horses have virtues lacking in tractors—they can learn a job rather quickly and perform their role automatically; they are often good company and have enough advantages that some people still use them profitably. They can also be balky—as this team was at the time—but tractors can be just as fussy, and nobody ever had to buy a new turbocharger for a horse.

Chapter 5

The Koster Ranch

M y friend Cliff Koster and his wife Onalee live on the farm Cliff's grandparents started 120 years ago, back in the early 1870s. The Koster place is in western San Joaquin county, very near the place where the grain combine was developed and refined; originally it was all in grain: wheat, barley, oats. Cliff's grandparents started the farm during the big California wheat boom of the post-civil war period.

The farm they homesteaded was part of what had been considered a great desert—the 500-mile long, 25-million acre Central Valley. This vast, ancient lake bed receives no rain from April or May until October and is a furnace during the summer. It was considered unfarmable by the early pioneers, but then they discovered how to till the virgin soil and suddenly, by 1870, California was giving up a different kind of gold. By 1869 the great central valley was producing about 16 million bushels of wheat.

Opposite, Bill Koster's great grandparents opened this land over 120 years ago, and built that barn behind him. While much of American and Canadian farming has changed, some families have managed to maintain the traditional way, despite urban encroachment, high taxes, undependable irrigation water, and grain prices that haven't changed much in seventy five-years.

Above, Cliff Koster's ranch has one of just about anything important to grain farming. This is an old header wagon, designed this way to maneuver under the aprons of a moving header-type harvester.

Next page, and this is what they looked like when they were in their prime; this big Holt combine still cuts wheat every year at Vista, California. *Robert Genet*

Holt Combined Harvester Model 32

Although the old Holt Company was officially merged into the firm that would be called Caterpillar, the Holt name and heritage lives on in Stockton, California, where it all began over a hundred years ago. Today Holt Bros. is a Caterpillar dealer, and when the opportunity to acquire an old Holt Model 32 harvester came along a few years ago, they knew they couldn't let it slip by; the company acquired it from the estate of a local farmer.

It was originally purchased by Edmund Uren, in 1928, a farmer with 176 acres in wheat and barley near the little Sacramento river delta town of Antioch. Like other western grain growers, Uren's operation was a dry farm, entirely dependent on seasonal rainfall for irrigation.

Edmund sold the Model 32 about 1941 to neighbors, brothers Archie and John Sullenger, also of Antioch. They used it for about ten more years, into the 1950s. Finally,

it was moved into a barn and forgotten. When the brothers finally passed away in 1989 the ranch passed into the hands of a nephew who was about to send the old Holt and all the residue of the farming operation off to the local dump so the land could be subdivided and developed. Harry Geddes, a Holt employee, heard about the harvester and the plans to scrap it; he had a chat with the current owners of the Holt company in

When the Holt Brothers company (a Stockton, California, Caterpillar dealer and direct descendant of the old Holt business) discov- *ered the remains of a Holt Model 32 harvester in a barn, waiting for the scrap heap, they acquired it and returned it to virtually* *new, working condition. And on a sunny September day, they hauled it back to the ground where it once worked.*

Stockton, Vic Wykoff and Ron Monroe, who both expressed an interest in the relic.

On May 8th, 1989, a team of Holt employees showed up at the old Sullenger farm with tools and trucks. The barn had to be disassembled, and the 64-year-old machine was removed. The rusty old side hill harvester was loaded on a lowboy trailer and lovingly transported back to the family, back to the Holt Company's big Cat dealership where the resurrection began. The Wisconsin gasoline 4-banger that provided power to the harvester was rebuilt, some of the wood needed replacement, fabric components were replaced and the entire thresher section overhauled. The rebuild took seven months.

There are a lot of old Holt combines still around in the California central valley; you see them everywhere, rotting alongside the road or piled in heaps, like the one at the big junkyard at the University of California's Davis campus. There are even a few in barns, collecting dust, waiting for another harvest. But the Holt Company had the foresight to restore at least one. And on a crisp September morning, the old Model 32 was trucked back out to a delta farm and hitched to a team of draft horses to have its portrait made for this book.

Above, Holt Model 32 combine detail: elevating mechanism. The design of the machine is an interesting combination of *heavy, cast iron components where strength is required, and lightweight wood and canvas elements where possible.*

Above, Holt Model 32 combine detail: the cylinder, apron, and beater are all visible here, where threshing begins.

Right, this piece of equipment is original to the machine and, although it was beyond restoration, was retained on the combine.

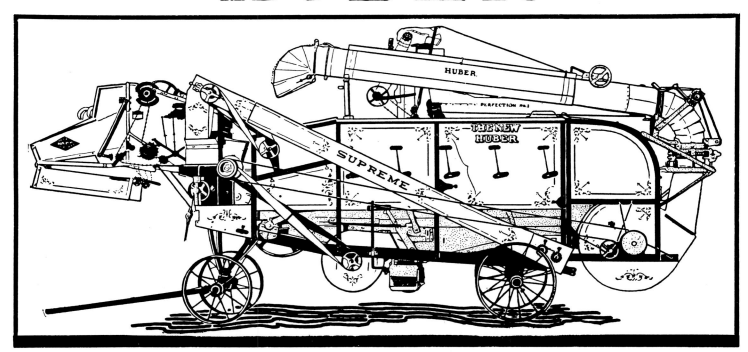

The Koster farm isn't a museum, but a visit gives you a tremendous sense of what grain farming must have been like during the glory days of a century ago. Part of that comes from Cliff's insights and recollections, and part comes from the Koster's interest in preserving their agricultural heritage.

The old Harris harvester is still in the barn, and it is still strong enough to bring in a crop, although it was retired in 1958. The mules that pulled the Harris were sold in 1914, but their harness and hitch are still neatly hung in the old redwood barns.

Two modern combines replaced the Harris, a Minneapolis-Moline from about 1950, and a John Deere from about 1955.

Cliff remembers harvest time during his childhood, before World War II:

"My dad rented one of the big Best steam harvesters back around 1910; with the steam coming from the tractor over to the harvester, the machine had almost unlimited power. He turned out 2,000 sacks a day on an 80-acre piece.

"The number of men on the crew varied. You had the mule skinner, or Cat skinner if you had a caterpillar tractor, the header tender (sometimes called the header puncher), the separator man—who would expect the most money, and one, two, or three sack sewers. Then on the ground you had at least two men 'bucking' sacks, that is, picking up the filled sacks from the ground and trucking them to a central location.

"The mule skinner was responsible for the team or the machine; he pulled the combine.

"The header tender was just about the lowest paying job; he raised and lowered the header, depending on the height of the grain, as well as helped with the servicing of the machine in the morning. It took about an hour every morning to go around and fill the grease cups with 'hard oil.'

"The separator man was the highest paid on the crew. He was the boss, and when things broke down it was his responsibility—with the help of the header tender—to fix them. The separator man showed up about a week before harvest to put our combine in good running shape before it went out in the field. That was my job later on.

"When I was the separator man, it was my custom to oil and turn down everything on the windward side of the machine. Our field was three miles around, and that took an hour and a half to get around, and I worked on the side of the combine that was out of the dust. We carried five gallons of crankcase oil and an old coffee pot; I put the oil in the pot and used it to slobber some on the drive chains, to lubricate them. Some people said this was a good idea, others said it was bad, but it made you feel better so I did it anyway.

"It took two sack sewers and a jig when you got into heavy grain, which wasn't very often. When you had a jig man, he bagged the grain and the sack sewers just sewed. Each of them could sew 700 sacks in a day. You knew exactly how many you did, too, because there was a counter on the combine that kept tally of each dump of grain sacks. The big Harris combine dumped eight sacks at a time, and the giant Best dumped ten. After every ten sack dumps, the sack sewers and the jig would change positions to provide every man with a little rest and variety from the routine.

"The sack sewer and sack jig's primary duty was to keep the sacks and twine organized. The twine came in big spools; you have to pull off a length and cut it, and to have plenty of cut twine bundled, ready to use. The sacks came in bundles of about 500, strapped together. At the start of the day each sack sewer had his own stack of sacks and bundle of cut twine right behind his seat. You kept your tobacco, extra needle, and sharpening stone all inside the seat box. The seat itself was bolted to the combine, and we padded them with old sacking to make them a little more comfortable.

"After the greasing was done, the sack sewers helped the header tender put fuel in the harvester auxiliary engine. We brought the fuel tank wagon over and used a long hose and a gear pump to refuel with white gasoline—white gas was a cent or two cheaper than the ethyl stuff.

"I can't tell you too much about the duties of the sack jigger because I don't think we ever hired one; when one was needed either my brother or I did the work, and we didn't get paid.

"What my brother and I enjoyed was listening to the tales of the sack sewers; they'd tell us a lot of tall tales, on the harvester, then later on after supper, too. They told about the places they'd been, and about how they had begun with the harvest, down around Tulare Lake, and how they would follow the harvest north, through the central valley, then up to Oregon, then Washington, and maybe up into Canada. In the off-season they just seemed to hang around Skid Row. They mostly seemed to be bachelors although some were married. Ordinarily we got different men each year, although once we ended up with the same sack sewers.

"I ran the header for the first time in 1938; I was sixteen then. It was still the depression and 1938 was the first good crop year we'd had in a long time. I helped on the next four harvests, until I went in the Army after the 1942 harvest. For two of those years I drove our 60 Caterpillar pulling the harvester, and the other two seasons I ran the header. I never worked as the separator man until after the war.

"I took the Harris harvester out in 1958 and was going to thrash with it. We set it up and started cutting the field behind the house; I had just gotten started admonishing the crew and getting them straightened out before they wrecked the machinery but before I could get them educated, we got plugged up and broke the first beater shaft.

"Now, behind the cylinder in the Harris are two beaters; they kind of paw through the straw and get it all aligned. The first beater is real important because it included the attachment for the pitman arm that drives the whole shoe on the back of the machine—and it broke right through the bearing. Well, I didn't have time to 'mechanic' the problem at the time, and my neighbors just bought a pair of John Deere model 95 harvesters, the first year they were offered for sale, with the idea they'd use them for custom harvesting, and I hired them to take over.

"We had heavy rains in April so '58 was a heavy crop year, and we got 25 sacks of bushel-weight barley an acre that year, and the John Deeres did it all. So 1957 was the last full season the Harris worked.

"Back then, before we started irrigating, we had 600 acres in barley. It took us about five or six days to bring it all in with the Harris.

"When do you harvest? That seems to depend on your level of paranoia. I know of some people around here who waited until the beginning of August when we normally harvest barley here at the last of May. Generally, we wouldn't wait much beyond about the 20th of May for barley, but there were people harvesting in August and getting away with it. And, one year, we needed some seed and in November we decided to cut a field that we had abandoned during the regular season. Now, that was interesting, and I had never done that before; practically everything was still there, and hardly any of it was shattered or busted out. It was full of weeds and it had been rained on—but it thrashed easy! I opened the cylinder way up because I figured I could hog it in because the beards and everything came off nice. It was nice, fat stuff, too; that was interesting."

Chapter 6

Corn Binders, Shuckers, and Shellers

There are plenty of tough, dirty chores on the farm and one of the toughest and dirtiest is harvesting corn the old way, with a knife. But corn is an extremely important crop, more valuable on the market, acre for acre and bushel for bushel, than wheat or barley or rice. It is the primary feed for the really big cash crop on the farm—cattle and hogs, and that makes it a very important crop indeed.

The plant is big, tough, heavy and resists the gentle treatment that so easily separates wheat and other cereal grains from their straw. You can whack a ripe corn plant with a baseball bat and the head will still not shatter; it might even whack you back. But, as it turns out, that is just as well since the entire corn plant makes great cow chow. While it has always been an important crop, until about 1890—when nearly all cereal grain harvesting was cut and threshed mechanically—corn was cut with a long knife, gathered in "hills" by hand, and shucked by hand.

McCormick started selling a corn binder in 1898, a machine based on the proven grain binder machine, but with a suitably heavier cutterbar and with a sturdy

Opposite, detail of a Circa 1900 Sandwich "Jupiter" corn sheller.

McCormick-Deering corn husker-shredder. This stationary harvester uses a system of interlocking rollers to snap the ears from the corn stalks—and it is about to be put to work bringing in yet another crop, about fifty years after it was sold. This machine works as well as new and continues to do the work it was designed to do, well over half a century before. *Bill Ackerman*

Above, International Harvester's horse-drawn corn binders reached their zenith with this machine from about 1935. Although this one is ground powered, some had gasoline engines installed to run the binder, giving the horses a lighter draft. This angle shows rather clearly the two upper gathering chains that guide the cut corn plants through the elevator section, to the binder unit.

Right, McCormick-Deering 4-roll corn shucker-shredder. Once your corn has been cut with the binder, shocked, and allowed to dry in the field, you bring it in to be shucked with this machine. This one is owned by Allen Smith who still uses it every year at the Farm Heritage Show near Chicago, and in Allen's own operation. It was built about 1929 and runs like brand new. *Bill Ackerman*

gathering mechanism built around two sets of chains. It was a one-row machine—and it transformed the harvest. You still had to gather up the bound shocks and assemble them in hills to dry; hard work, for sure, but the mechanization process, to the relief of many of the hired hands and farm boys who normally were tasked with the chore, had begun.

Within a few years the corn binder was a widely accepted and common tool on the farm. The early ones were pulled by one or two horses (normally two—it is hard pulling), and the machines were driven by the same type of ground wheel used on mowers and reapers. But ground drive and horses both have some drawbacks: the first requires dry soil and good traction to work properly, and the second have notoriously poor fuel economy under normal operating conditions.

But the simple horse drawn, ground powered corn binders were sold by the tens of thousands to grateful farmers—and some of these horse-drawn machines are still very much in use. I found one at work on a Sep-

tember afternoon on an Amish farm outside Jamesport, Missouri, a nice old International Harvester machine astern a pair of handsome Belgians. The binder clattered musically as it marched down the rows, one by one, plucking the corn stalks, gathering them up, then dropping them in bundles. One thing you notice right away when horses provide the horsepower—the silence, except for the machine, the rustle of the corn, and the swish-thunk of the big bundles as they topple to the ground.

He stopped to chat and I asked permission to photograph his binder for the book, knowing that he would certainly not wish to be included. He and the horses rested for a few moments while their visitor harvested the opportunity of the moment. To Amish and Mennonite farmers in Canada and the US, these machines don't belong in museums, collections, or displays, they are fine tools and important implements that belong in the field, behind horses, and they are right.

The Corn Binder

Corn binders have three basic components: a sturdy cutting assembly, the elevating unit, and the binding mechanism. The cutter knife assembly uses a pair of fixed, stationary knives ahead of the sickle; these help align and cut the heavy corn stalks. They must be kept extremely sharp and get regular attention during harvest. The sickle itself requires much more careful attention and adjustment than a cereal grain reaper, with all of the looseness from wear adjusted out; properly done, with sharp blades, the result is a light draft and easy pulling for the team.

Above, Circa 1900 Sandwich "Jupiter" corn sheller on display at Midwest Old Thresher's Reunion.

Next page, the Amish farmer who owns the binder and the team of handsome Belgians kindly consented to the photography of the rig, but not of himself and is taking a break out of the shot. Such farmers manage to use technologies most farmers consider hopelessly inefficient and obsolete—and become prosperous and secure in the process. Their horses are sometimes as sleek and handsome as any Ferrari.

Three pairs of chains gather the stalks, keeping each upright, and delivering them to the binding unit. One pair is mounted low, at the front, and the others are mounted at the middle and upper portion of the gathering boards. All are synchronized and

While most of the machines on display at Mid-west Old Threshers Reunions are tractors, a few of the tractors are embellished with harvesting equipment, like this John Deere corn harvester.

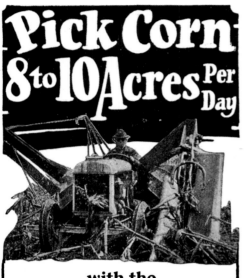

Pick Corn
8 to 10 Acres Per Day

with the
CONTINENTAL CORN PICKER

Now you can pick corn easier and cheaper with Fordson power. You can do a better job at lower cost with this simple, sturdy, reliable Corn Picker built especially

for the FORDSON

Mounted on the Fordson — carried by it, not pulled. Get into the field and pick in any kind of weather — go anywhere the tractor can go. Built for tough service — yet 1000 pounds lighter than any horse drawn Picker. Attached or detached in a half hour — on and off as a unit. Four years thorough test proves it does its work well under the most difficult conditions. A godsend to corn growers. Fully guaranteed. Write for information.

Continental Cultor Co.
Dept. 182 Springfield, Ohio
Manufacturers of the famous CULTOR
— the Power Team for cultivation and
other Light Draft Field Work.

keep the corn plants in a vertical embrace until each is released to the binder.

As with a grain binder, the corn binder is equipped with controls to let the driver adjust the machine for best results. Cutter height, balance, relative ground tilt attitude, and butt pan height are all under the control of the driver.

Finally, the power bundle carrier is tripped with a foot pedal, as with the grain binder, kicking the bundles off the machine for pickup.

John Deere Corn Combine Model 10

Deere introduced a corn head conversion for combine use with their Model 10, a 1954 release. A beefier combine, designed specifically for the rigors of corn harvesting, was called the Model 45, a self-propelled combine with 26in cylinder and 8ft, 10ft, or 12ft header. Although the 45 superficially resembles the 55 externally, the corn harvester uses heavier components and stronger fabrication techniques than the earlier machines designed for work in grain.

This machine is called a "sweep" and allows stationary threshing machines to be powered by teams of horses or mules.

Two hundred and ten bushels of corn pours into
the truck from a Case 1688 combine. *J.I.Case*

Above, a corn head attachment was an option for the John Deere model 45 combine, shown here in a 1954 photograph. *John Deere*

Left, Circa 1900 Sandwich "Jupiter" corn sheller on display at Midwest Old Thresher's Reunion. Chain drive was a recently proven technology when this machine was designed, with tremendous advantages over gears (which are also extensively used on this same machine). Chains provide the same capability for synchronous drive, but with less friction; and, by shifting the chains from the inboard to the outboard sprockets, the operator can quickly change the processing speed of the machine. Also, when a link fails it is a simple matter to install a new one; adding a new tooth to a gear is a more complicated problem.

Chapter 7

Tomatoes, Cotton, Cranberries—
the Odd and Exotic

Better Farm Buildings

Conducted by **R.D. Radford**

Mechanized tomato picking is a rather recent development. Until about twenty years ago the skin of tomatoes suitable for commercial growing was just too thin and fragile to stand up to the trauma of mechanized harvesting. That changed when botanists and farm technology specialists at the University of California at Davis teamed up to develop a new variety of tomato and a new machine to pick it. The result was a tomato tough enough to play ball with, with a very high "paste" yield, and that helped create a new agricultural industry. Now we can all buy canned tomatoes at extremely low cost.

The early machines were very slow, extremely large, and needed between sixteen to twenty people aboard as "hand-sorters". The machines traveled about half a mile per hour—when they were blasting along full tilt—with about twenty people trying to keep up the crop. Most of those people were removing the green fruit, dirt clods, and pieces of vine.

The basic process involves harvest of the whole plant, with roots and dirt clods often included. The vines are fed into a rack or

Opposite, an aerial view of tomato harvest in full flood. The season lasts about two months, September and October, during which time the canneries operate around the clock.

table section where they are vigorously shaken to loosen the fruit and to knock off most of the dirt. Early machines left a lot of good tomatoes on the ground and kept a fair amount of dirt and vine to be hand sorted from the crop. It was a process that was only partly mechanized.

Yields at that time were about fifteen to twenty tons per acre. Then electronic color

About the only time you'll see a tomato harvester this clean is when it is brand new because tomato harvest time can be a bit muddy—or juicy. But this is a radical piece of technology, the result of pioneering work at UC Davis about thirty years ago, that combined a new kind of tomato and a machine that could harvest it. This is the second-generation design, requiring a crew of about nine people instead of the original twelve.

Above, the only reason we can buy canned tomatoes, tomato sauce, and tomato paste at the low costs currently available is this machine and the very special tomatoes it is sucking into its gaping maw. The UC-Blackwelder tomato harvester seen at work on Ron Del Carlo's field here needed a sturdy fruit that could endure the rigors of mechanized picking, sorting, and handling. That tomato only was developed about 1970.

Right, although these tomatoes might not have all the taste of other varieties, they tolerate this kind of handling while traditional varieties would turn automatically to sauce—or sludge.

sorters came along and that increased the capacity of the machine substantially, at the same time reducing the number of people that needed to be aboard. Suddenly, the harvesters were going two or three miles an hour.

Plant breeders were unable to take their research to its logical conclusion, the organic baseball, but they were able to produce

Above, back to the future, Part C—where the only way to go forward is in reverse. This elderly International Harvester 320 single-row cotton picker was part of the first generation of really practical machines and had a lot to do with the rapid expansion of the industry. The picker is designed to be removable, but it is so complicated that it would take most of the growing season to get it back on or off.

bigger yields. By about 1980, each acre produced around twenty-five to thirty tons of tomatoes, some of which actually had some taste to them, and all of which tolerated the violent shaking of the harvester. With increased yields, faster machine speeds, and improved harvesting technologies, the industry quickly matured.

During the early 1980s, FMC Corp. helped develop the brush shaker—a large device shaped like a big bottle brush—that vibrates and ro-

Next page, while much of the business of tomato harvest is mechanized, it still takes about ten people to crew the machine—to discard the dirt clods, the damaged fruit, and the bits of vine the machine misses.

The tomato harvester moves slowly, ponderously down the rows, cuts the whole plant off at the ground and separates the fruit from the vines with a series of rubber fingers and electronic sensors.

Left and above, the CPCSD's International Harvester 320 cruises up and down the rows all day and the bin still never fills up. That's because all the cotton goes into specials sacks, each carefully tagged and weighed immediately after harvest. Each row in this breeding nursery contains a separate hybrid, part of the continuing development of better cotton.

tates, removing nearly all the fruit from the vine with very little loss. All contemporary manufacturers—Blackwelder, FMC, Johnson, Pickrite, and some European players—use the same basic brush-shaker technology.

California alone now harvests ten million tons of canning tomatoes, each ton worth about $50 to the grower—a half-billion dollar industry in one state alone.

Cotton

When most Americans think about farming and field crops, they think about wheat waving in the wind—but folks in agriculture know better; cotton and corn are where the big bucks are. Here in California cotton is usually the #1 row crop (competing with grapes) in dollar value. "Most people, when they think about cotton, think about the South, but that is not where this product is produced any more," says Tom Cherry, president of CPCSD. "Texas and California together produce about half the nation's cotton crop. California alone plants about 1.2 to 1.4 million acres in cotton and produces about 2.5 to 3.5 million bales every year. Texas plants about 5 to 6 million acres and produces about 3.5 to 4 million bales."

When mechanization came to harvest time, the inventors started with the easy

Left, the spindles engulf the cotton plant, enmeshing it and harvesting only the mature cotton while leaving nearly all the other plant material. Clean, uncontaminated lint depends entirely on the proper adjustment of these spindle assemblies; there is no cleaning section on a cotton harvester, unlike grain combines.

Right, no, it isn't a lawn mower in need of sharpening—it is a cranberry picker. The horizontal bars break the berries loose from the vines, then gathers them in a sack at the rear of the machine. *University of Massachusetts—Cranberry Experiment Station*

Lower left, these spindles rotate only when in contact with the plant. Small barbs on each snag the cotton "lint" from any bolls that are ripe and open and the cotton is snatched out, to be collected by the assembly on the right.

crops and saved the tough ones for later. Cotton was one of the tough ones. The fibers are held in brittle bolls that ripen unevenly, from the top of the plant down. The grade of the crop and the price it will command depends partially on its condition, and until fairly recently the only way to keep it free of most "trash" and contamination was to pick it by hand. That changed only after World War II when crops like tomatoes and cotton—previously hand harvested by necessity—could finally be picked mechanically. Major manufacturers like Allis-Chalmers and International Harvester started selling their first commercial models about 1950, but that was only after many years of development and expense.

It wasn't that people hadn't tried; grain kernels want to break out of the straw but cotton holds tenaciously. Many hundreds of patents were issued for devices intended to mechanize the process, some using vacuums, others using static electricity, while others simply ripped the whole boll off the plant, intact. The concept of the spindle stripper was well developed by the mid-1920s and was tested by International Harvester and other implement manufacturers, but, needed another fifteen years of development to fully ripen.

International Harvester Company began a program to develop a commercial machine in 1901, but it was not until forty years and over $5 million later that they had a functional machine. The technique developed uses two sets of spindles, one rubber and the other steel. As the picker travels down the row, these spindles pivot to meet the plant, then mesh through it in a kind of mechanical embrace. The spindles each spin on their own axis, and when they come in contact with the exposed cotton fibers of a ripe boll, the fibers tend to be grabbed by the spindles and sucked right out of the boll, the "lint" fibers wrapping around the spinning spindles. As the picker moves up the row, the spindles continue to rotate, finally disengaging from the plant—then the spin-

dles stop their spinning. Finally, the cotton—with the seed still embedded—is gently removed and blown into the storage bin.

It sounds complicated—and it is. The first machines weren't very good, and there was a lot of initial resistance to mechanical picking. There was too much trash and contamination from mechanically picked cotton; besides, just as in most other commodity industries since 1750 or so, the hired help objected to the prospect of unemployment. One Alabama cotton grower, I recall, bought an early version of the mechanical picker back in the early 1950s that turned out to not work very well; rather than sell it, he parked it in plain view of the hand pickers as a kind of motivational device or subtle warning.

In the grand old tradition, early pickers were all tractor mounted with some provision for unbolting the tractor for use during the rest of the year. Many farmers, however, decided it was too durn much work to get the two machines untwined and simply dedicated a tractor to the cotton harvesting machine.

International Harvester's early commit-

Right, here's an early model of a cranberry bog water reel harvester; this machine is used while the bogs are flooded and merely knocks the berries loose from the vine. *University of Massachusetts—Cranberry Experiment Station*

ment and long dedication to the cotton market turned into a long line of successful machines. The first IHC commercial pickers appeared during World War II, a small run of only about seventy Model H-10-H harvesters based on the Farmall Model H tractor sold during 1944 and 1945. As with most cotton pickers, it is difficult to tell if it is coming or going—the tractor is reversed under the machine and it moves forward while in reverse gear. The Model M-11-H replaced the 10 in 1946, again with quite limited commercial success since only about

Below, once the berries are released from the vines they are collected and processed. *University of Massachusetts—Cranberry Experiment Station*

rather costly but each replaced about forty farm laborers at harvest time, and they helped convert the cotton production business from an industry of small-time operators with few acres under cultivation to one characterized by huge operations with vast acreages in production. Even the early models like International Harvester's M-12 (built from 1946 to 1952), a one-row/tractor-mounted machine, could harvest about twenty acres a day, picking ten bales worth of seed cotton.

International Harvester introduced the Model 314 and 320 in 1961; both were one-row machines, both mounted on Farmall 560 mechanical transmission tractors, but the 314 was set up as a single fan, low drum unit while the 320 used a high drum configuration. Two row machines came along four seasons later, in 1965, and diesel power in 1968. Then, in 1978, International Harvester finally put the operator in a glass box, just like the other manufacturers: a pressurized cockpit with stereo sound, air conditioning and everything except autopilot.

International Harvester 502

There are times and places where new and improved just doesn't work better, and one of those places is in the breeding nursery where seed cotton is developed. California Planting Cotton Seed Distributors (CPCSD) develops and markets improved cotton seed, mostly for the California market but also to the rest of the world. A large part of the business involves breeding new or improved varieties. When it comes time to harvest the dozens of different varieties of cotton in the CPCSD nurseries, a rather unusual system is required—an elderly, simple, slightly customized 1960 International Harvester cotton harvester mounted on a 1950 Farmall tractor.

The old Farmall and McCormick-International Harvester Model 502 single-row picker happen to be ideal for CPCSD's specialized harvesting situation. Each 50ft row of cotton represents a specialized variety and must be individually harvested. While nearly all the other IHC 502s are long since retired and rusty, this thirty-year-old machine, customized for the specialized needs of the breeder, is likely to soldier on well into the next century.

sixty-five were sold. But the domestic cotton market grew rapidly during the 1940s and production was up to 15 million bales by 1951, over three times the harvest of a decade earlier—and the price of a cotton picker was down to about one half the cost of the first models.

Allis-Chalmers, first machine appeared in 1949, a two-row picker based on the company's WD tractor. This pioneer model for the company could be adjusted for row spacings from 36in to 42in and held up to 1,200lbs of cotton.

Each of these machines may have been

Above, the Case regular peanut thresher was available in two sizes, 20x28 and 22x36 and could be equipped with either hand feed or self feed. Inexpensive attachments could be purchased to adapt the machine for harvesting various types of beans, as well as husking and shelling corn. *J.I. Case*

Above, the special cylinder and concaves for the Case peanut thresher. *J.I. Case*

Opposite, another early mechanical water reel harvester in action, about 1970. *University of Massachusetts—Cranberry Experiment Station.*

Chapter 8

The Shrinking Marketplace

Agricultural harvesting equipment was once one of the dominant industries in Canada and the US, but, in a way, the success of mechanized farming killed off not only the small family farm, but the very industry that enabled that success. Of the dozens of major (and hundreds of minor) manufacturers of harvesting equipment that existed around 1900 only a handful remain, the rest swallowed up or run out of business by a marketplace that has changed dramatically. As an example, consider the International Harvester Company's history and evolution from a conglomerate around the beginning of the century to just another component of one of the survivors today. Its history and its product line represent just one of the many stories behind the evolution of harvesting machines in the US and Canada.

Opposite, John Tower and daughter take the old John Deere Model 55 out to try to bring in the wheat, despite a weedy year. This ground, in the Gold Rush foothills of California near the tiny village of Copperopolis, has belonged to the Towers since 1852; the John Deere 55 came along about a century later.

The old Massey-Ferguson 750s on the Harper ranch don't have all the bells and whistles of the new machines but they have one important advantage over the combines in front of the dealer's showroom downtown—they're paid for. With reasonable care and regular maintenance they are still going strong after twenty seasons and show no indication of quitting just yet.

International Harvester

Although International Harvester Company's combines are now part of the J. I. Case lineup, the heritage of this company accounts for a big part of North American agricultural history and many of the pioneer machines to bring in the crops wore an IHC logo.

IHC was born in 1902, the product of five parents: McCormick Harvesting Company, the granddaddy of all American manufacturers, plus
• Milwaukee Harvesting Company

Opposite, it was a good season for wheat on Allan Harper's ranch outside Sundance, Wyoming, with good moisture toward the end; the crop would have made about fifty-five bushels to the acre, but a late hailstorm knocked about a third of it out of the heads, but it still made about thirty-five—not bad, considering. Allan is a re-formed attorney now practicing farming on his family's 6,000-acre operation, with a thousand of those acres in grain. Here he comes now in one of the ranch's old 1975 MF 750s.

• Marsh Harvester Company/Deering Harvester
• Warder, Bushnell and Glessner Company
• Piano Manufacturing Company
Then, a year later the International Harvester Company grew again, adding
• D. M. Osborne Company
• Minneapolis Harvester Company
• Aultman-Miller Company
• Weber Wagon Works
• Parlin and Orendorff Company

The result was a new company that quite suddenly dominated the American harvester market and became a virtual monopoly. International Harvester had dealers in just about every crossroad village and up to five dealers in many farm towns and cities. Despite the presence of such powerhouse companies as John Deere in the market, International Harvester owned eighty-five percent of all farm machinery sales in the US during the first few years. Over the years this domination was eroded by court decisions, but even in 1927 IHC still sold about two thirds of all implements in the US.

A big John Deere combine rumbles through a dusty field of safflower, late in September. Modern combines are easily adjusted to thresh grains of very different size: wheat, oats, rice, safflower, corn—but they still won't do tomatoes.

No. 1 Harvester-Thresher

International Harvester's first actual grain combine didn't hit the market, though, until 1914, over a decade after the formation of the giant company. It was called the Deer-

Next page, the Allis-Chalmers company got reaped and threshed in one of the many mergers and acquisitions of the 1980s and is now a component of AGCO, the heir of the old Baldwin "Gleaner" line. This Model L combine is sucking up part of California's rather small wheat harvest on a hot June afternoon. California was the leading producer of wheat in 1880, but that crop has long since been replaced by cotton, grapes, and other high-value commodities.

107

With the advent of large-scale irrigation projects, rice now grows on land that was once a desert. Here's a John Deere harvester combining the crop as seen from about 1,000ft. Note the wide treads used in place of tires, necessary because of the sloppy going in fields that are likely to still be damp.

ing No. 1 Harvester-Thresher, a simple, small, pull-type, ground-drive machine with a 36in cylinder and a 9ft header.

With slight modifications, this same basic design stayed in production for another decade, although the company experi-

mented with modern technology, an auxiliary engine was added in 1915. This model, the Deering No. 2, was already intended primarily for use with a tractor, although you could still get it rigged for use behind a team as an optional extra; other options available

included a 3ft header extension, plus modifications that allowed you to use it as a stationary thresher instead of as a combine.

A hillside combine, called the McCormick-Deering No. 7 Harvester-Thresher, came along in time for the 1926 harvest. With a 12ft header

(plus a 4ft optional extension), the No. 7 certainly couldn't compete with the giant Holts or Harris combines with their 30ft to 50ft cuts, but it allowed some grain farmers to harvest wheat, barley, oats, and other grain crops on land that would otherwise be fit only for pasture. This machine could be leveled on 65-percent slopes; it stayed in production until 1932 and sold well in the US, Canada, and abroad.

Machines of this type were the only output from International Harvester right through World War II—simple, economical, compact pull combines with 24in cylinders, mostly, and headers from 5ft to 16ft. You could still buy a ground drive machine from International Harvester as late as 1951, but after the war even conservative old International Harvester started offering self-propelled combines.

International Harvester 123SP

Although the very first self-propelled grain combines appeared well before World War II, back in 1938, it took another fifteen years for International Harvester to offer one, the 123SP. This little combine brought in the 1942 harvest and looked suspiciously

like the earlier No. 62, a pull-type machine that appeared a year earlier. The 123 was powered by a six-cylinder IHC engine, weighed a bit over 7,000lbs, used a 12ft header, and was available with all the options: pickup attachment, bagger, straw spreader, and—for the sissies—a parasol. About 11,000 were sold during the six-year production life.

IHC Self-Leveling Combine No. 141H

Prior to 1955 and the introduction of the 141, International Harvester's combine products tended to be extremely safe and conservative designs based on principals and patents developed by the competition. But the 141 (still a pull-type with PTO drive) used a rather radical hydromechanical system to provide, for the first time, automatic four-way leveling. This machine, with a 16ft or 18ft header, maintained level fore-and-aft/left-right all by itself. The trick was a 100lb pendulum mechanically coupled to a

Right, like most combines, the 55 is a working machine, not a collector's item, so strict authenticity is not a requirement for operation. When it was time to service what John Tower calls "early combine air conditioning," the air conditioner came from Deutz-Allis.

system of hydraulic actuators; the body of the machine stays level on ground that slopes up to 16 degrees left or right, 6 degrees nose down, 18 degrees nose up.

Hydrostatic Drive, Axial Flow, and Space-Age Technology

International Harvester introduced another important development nine years later, in 1964, with the first hydrostatic drive on an SP combine, available for models 303, 403, and 503, and within several years manual transmissions weren't even an option anymore.

About this time American and Canadian farmers started getting all soft and flabby; International Harvester, along with all the other manufacturers, started providing enclosed cabs with air conditioning, sound-proofing, and stereo sound. The 914 model, first sold in 1970, began the trend that has made combine cockpits start to look like something out of Star Wars, what with all the monitors and digital readouts.

Then, in a really radical maneuver, International Harvester introduced a completely new kind of separator and cleaning system with the Model 1440 combine that was first sold in 1977. This machine, and the others of its generation, uses a large, long, tubular threshing system that treats the crop in an entirely different way than the old technique. International Harvester called this new method "axial flow." Instead of many small, vibrating, intricate, fragile parts, this new system uses one large rotor, about 24in in diameter, through which the cut crop migrates.

John Deere

Another example is the John Deere company, a relative latecomer to the combine business, but another of the survivors. The "green machine" has dominated the cultivator implement market since 1837, but only got into the harvesting machinery business in a serious way about 1907. The John Deere company's first effort in the market was a twine grain binder; seven experimental

Left, here's what John Tower's old John Deere model 55 looked like when it was new. The photograph dates from 1950. *John Deere*

Opposite, a flotilla of modern combines sweep across the grain fields of the American midwest. Machines like these cost about $200,000 each, but pay their way with extreme efficiency and speed of harvest. Many are purchased by custom harvest contractors who typically use them for just one season, while virtually all repairs are still under warranty, and then trade them in for the new model next year. *J.I. Case*

After a circuit around the field, John had some wheat in the bin, but he was ready to put a match to the field because of the weeds. Since rain in the summer isn't a problem in his area, he let the crop stand for another month or two, till the weeds dried enough for the combine to handle without plugging. The crop is destined to feed his wife's flock of sheep.

versions were assembled in 1910 and tested that summer. One of the designs was put into production with a run of 500 machines built in a rented factory owned by a competing implement company, the Marseilles Manufacturing Company.

In 1911 John Deere Company acquired Marseilles Manufacturing, the Moline Wagon Company, the Kemp, Dain and Van Buren Manufacturing companies, and in 1912, the Syracuse Chilled Plow Company. These acquisitions made John Deere Company an agricultural version of General Motors, with a broad product line, extensive dealer network, and tremendous resources. Perhaps as a result, the company sold 2,000 of the new grain binder in 1912, and began

to produce harvesting machines by the thousands—mowers, reapers, binders for grain and corn, but no combines; those were still conceded to the competition.

John Deere & Company's First Combine

Big Green finally decided in 1924 to offer a grain combine and two designs were developed, one with a 12ft platform (called Model 1) and a second with a 16ft platform (Model 2). It took the company two years to develop the machines to maturity and bring them to market; Model 2 was first, with initial sales in 1927, Model 1 sold for the first time the next season, 1928. Both used a 24in cylinder, 60-bushel grain capacity, and both were very conventional pull-type machines intended to be used with horses and ground-powered through a bull wheel. Model 1 was offered with either an 8ft or 10ft header while Model 2 came with a 12ft or 16ft platform. Both were promptly replaced with slightly improved versions, the Model 3 (30in cylinder) and the Model 5 (24in cylinder).

During the Depression years the farm implement market was severely affected; few farmers had money for new machines, and many parts of the US and Canada experienced successive years of drought and crop failure between 1930 and 1938. John Deere Company continued to produce small, compact, conventional, and affordable combines—including the tiny Model 10A with a 3.5ft sickle bar.

Auxiliary engines became available for the first time on John Deere Company machines in 1936 with the Model 6, a petite little 6ft platform combine. Model 9, first sold in 1939, provided PTO (power takeoff) drive for the first time on a John Deere combine, and also replaced the canvas apron with an auger for the first time.

The Model 12A was introduced that same year and would remain in production until 1952, with over 116,000 being sold. The 12A was a small, economical combine that seemed just right for operations with-

Opposite, John Deere 55 header and reel system.

Cliff Koster couldn't figure out why the cleaning shoe on his old John Deere combine wasn't working right and wanted a better look; that's the windshield from an old Ford car that he installed in the machine to watch it at work . . . but he never did quite figure out what the problem was.

Opposite, Bill Koster coaxes the 45-year-old Minneapolis-Moline into action—its annual moment of glory.

Below, Cliff Koster's Minneapolis-Moline has pretty well paid off the bank loan by now after around forty-five seasons of bringing in the grain. Koster's previous combine is still in the barn—a 1914 Harris, originally pulled by mules. The harness for the mules are still in the barn, too.

This fine old Baldwin "Gleaner" combine is one of the few of its ancient breed still alive, and is on display at Midwest Old Threshers Reunion. Baldwin was once a major manufacturer of combines, specializing in the smaller models like this one, suitable for family farms with diversified operations. Motive power is supplied by a fine old Fordson tractor.

out large acreage in grain; it was available with a 6ft or 7ft platform and a 60in cylinder, powered by an auxiliary engine.

John Deere Company bought out the Harvester Division of the Caterpillar Tractor Company in 1936, and with it the experience and insight of those two ancient and honorable combine builders, Holt and Best Manufacturing companies. Only one of Cat's designs was preserved intact, the Model 36, a hillside harvester with 22in cylinder and either a 16 1/2ft or 20ft cut, but it must have been a good one because it stayed in production until 1951.

John Deere's First SP Combine

World War II prevented much development in agricultural harvesting technology; most farmers and farm implement manufacturers made do with what they had. But self-propelled grain harvesters had been proven by other manufacturers and in 1945 John Deere started a development program for such a combine. That program matured two years later with the highly successful Model 55SP, a sturdy little combine with a 30in cylinder and your choice of 12ft or 14ft header. More than 80,000 were built before Deere ceased production of the 55 in 1969.

One Model 55 is still in use on the Tower ranch outside the tiny village of Copperopolis, California, where I watched John Tower use his to harvest his special blend of wheat and weed seed after an uncooperative growing season. Although John enjoys and restores antique farm machines (and owns a fully functional Advance steam tractor), the Model 55 is strictly for go, not show. It was purchased nearly new by John's dad about 1968 from a neighbor.

The Tower family started this ranch in 1852 and a hundred years ago were growing a lot of wheat and barley, like many other California farmers of the time. Only about 100 acres are still in grain now, oats and barley, used for feed for the family's small flock of sheep. After John gets the grain in he gets out the old Alliance steam tractor and belts it up to the rolling mill to process the feed for the sheep.

Right, Detail, Baldwin "Gleaner" combine.

Below, Minneapolis-Moline logo.

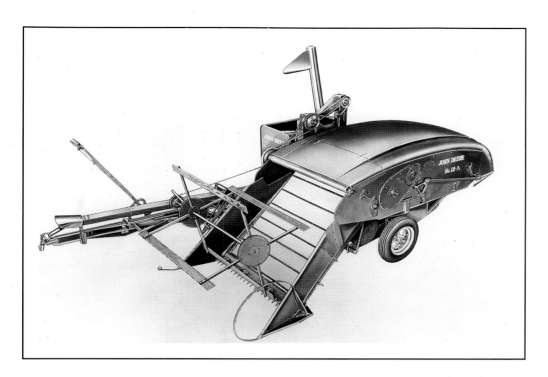

Left, the tiny John Deere Model 12 combine was just the ticket for many New England farmers with small acreages and irregularly shaped fields. It was popular in the early 1950s when this shot was made for promotional applications. *John Deere*

Below, Detail, Baldwin "Gleaner" combine.

Harvest Time at Ardenwood Historic Farm

When John Horner pioneered grain combining here in California back in 1852, the man he hired to build his machine was James E. Patterson. Patterson ultimately became a wealthy farmer whose operation included more than 6,000 acres under cultivation and in pasture at the eastern edge of San Francisco Bay, near the little rural village of Newark. Patterson called his lovely farm Ardenwood, and it thrived for over a century.

But after World War II, California was engulfed by a horde of humanity and the little village of Newark quickly became a town, then a city, and the fields and pastures disappeared under houses and shopping malls. The Ardenwood farmlands supplied much of the acreage. During the 1970s, though, the Alameda County Board of Supervisors were able to save the last 205 acres of the Patterson homestead. It is now one of many historic, "living museum" farms that preserve some of our agricultural heritage—a place where you can watch people cut grain with a cradle and with an old International

Harvester binder pulled by horses—a place where you can still learn to do both yourself.

A visit to Ardenwood is a bit like visiting 1903 or maybe 1930. A big Best steam tractor shares the barn with a Case thresher and a small herd of ancient farm trucks and machines. But these aren't dusty displays of dead machines with rope barriers around them and signs saying don't touch. All of them work and all of them are used to do the same work for which they were designed.

Ardenwood plants wheat and corn in the spring in ground that has been tilled and planted with horse-drawn equipment. The wheat is cut with a horse-drawn binder and fed into the old Case thresher. The corn is cut by hand and shelled with hand-cranked shellers. But unlike in James Pattersons' time, farm labor isn't a problem; families come by the hundreds to participate. School children who've never been outside a city help pull carrots—and are amazed that they come out of the ground dirty. These kids get to experience the size, power, and utility of Ardenwood's fleet of fourteen big draft horses—Shires, Belgians, Percherons, Clydesdales—and several old working tractors.

About 55,000 people visit Ardenwood every year, and thousands of others visit similar "living museum" farms in other parts of the US and Canada. While it may be true that those days are gone, they certainly

aren't forgotten. Says Ardenwood's curator, Ira Bletz, "people still care about our farm heritage, and there are still people who have the vision to preserve it.

"Of all our special events, our threshing bee is the most historically appropriate. We harvest wheat at the same time that people used to harvest wheat on this land, using the same methods and machinery. But our urban and suburban visitors can't seem to understand what threshing is. They seem to understand and appreciate how and when we harvest walnuts and tomatoes; October is our busiest month. Everybody thinks of farms at harvest time. But they just aren't familiar with threshing.

"Now we are learning how to involve and educate people in things like threshing and that is our biggest problem. How do you take something like the threshing machine—something with belts and wheels and all kinds of safety hazards—and let visitors without any training or experience with such a machine, do more than just watch it in operation . . . how do you let them participate in using it? We have, over the years,

Left, Ardenwood's Dave Cook sets up the John Deere hay rake. While not as complicated or glamorous, forage crop harvesters were (and still are) essential elements of operations with significant numbers of livestock.

Next page, while Ardenwood doesn't let everybody drive the teams or run the thresher, anybody can toss bundles in the wagon—or lean on a pitchfork—in the ancient traditional ways.

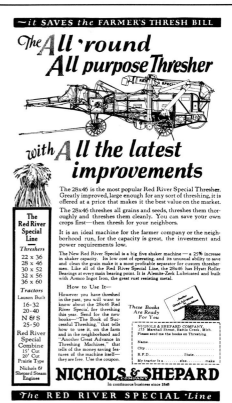

Above, you could still buy a thresher like Ardenwood's as late as the 1950s, and about 300 farmers bought such machines from Case alone in 1950.

found ways of helping people get the feeling they are actually involved in the process—without having to worry about getting them sucked into the machinery.

"We give people something of the feeling of what it was like at harvest time on a farm back around 1890 to about 1920. When you come in here, you leave the late 20th century; we are an island of 19th century farming surrounded by 20th century homes and businesses. We let 300 or 400 people go out into our corn field to harvest corn—it is an experience they can't get anywhere else. When you go into the corn field you loose the crowd, the activity, the 1990s, the freeway, the houses around our farm, they all disappear and you are surrounded by corn. You can't see the houses or the freeway or anything that doesn't look or sound as it would have if you were in a corn field 100 years ago. Kids and families go in there and it is a wonderful experience for them! They are amazed that we permit them to actually have the experience of harvesting this corn, and for some it has become a family tradition. We've been able to get people to do almost everything here—to plant and harvest, and even to shovel manure—the only thing we can't get them to do is to weed."

Ardenwood Historic Farm
34600 Ardenwood Blvd.
Fremont, California 94555
(510) 796-0663
Open from April through November, Thursday through Sunday, 10 to 5.

Appendix

Clubs, Associations, and Events
Recommended Reading

All Tractors
Antique Power
Patrick W. Ertel, PO Box 838, Yellow Springs, Ohio 45387

Allis-Chalmers
Old Allis News
A quarterly publication with a bit of everything A-C ever built. Subscriptions from Nan Jones, Editor, 10925 Love Road, Bellevue, Michigan 49021; $12 per year.

Ferguson
The Ferguson Journal
George Field, Sutton House, Tenbury Wells, Worcestershire, WR15 8RJ, England

Ford
The 9N-2N-8N Newsletter
Gerard W. Rinaldi, 154 Blackwood Lane, Stamford, Connecticut 06903-4707

International Harvester, McCormick-Deering
Red Power
Daryl Miller, Box 277, Battle Creek, Iowa 51006

J.I. Case
Old Abe's News
Case Collector's Association Inc., 4004 Coal Valley Road, Vinton, Ohio 45686
The association publishes a lively, entertaining magazine and sponsors two big shows each year. Although most of the member interest is concentrated on tractors, there is still substantial interest in collecting and restoring harvesting equipment. The association's magazine is edited by David Erb, and is definitely worth the price of membership.

John Deere
Green Magazine
John Deere fans have an excellent resource in the very slick Green Magazine, available from Richard Hain, editor, RR1, Box 7, Bee, Nebraska 68314.

Massey-Harris, Ferguson, Wallis
Wild Harvest
Keith Oltrogge, 1010 South Powell, Box 529, Denver, Iowa 50622

Minneapolis-Moline
M-M Corresponder
Roger Mohr, Rt #1, Box 153, Vail, Iowa 51465

Minneapolis-Moline
Prairie Gold Rush
Roger Baumgartner, Rt #1, Walnut, Iowa 61376

Oliver
The Oliver Collector's News
Available through Turtle River Toy News, Dennis Gerszewski, RR1, Manvel, North Dakota 58256

Steam and Gas Show Directory
This publication lists hundreds of shows and events around the US and Canada, all devoted to the preservation of agricultural technology heritage. While many of the shows specialize in one kind of technology or another—steam engines, farm tractors, stationary one-cylinder gas engines, peanut harvesters, or stationary threshers—nearly all will include a bit of everything. The directory lists nearly all, often with additional ads providing detailed information.
These shows typically last two or three days, sometimes longer, and seem to be designed to be as much fun for the exhibitors as for the audience. Most have parades, contests, demonstrations, and plenty of live music—provided by chuffing old steam tractors, antique kerosene-powered farm tractors, or from odd little gas engines sputtering along at minimum rpm. Camping facilities (normally primitive) are usually available on site.
The directory is available from Stemgas Publishing Company, PO Box 328 Lancaster, Pennsylvania 17608. The 1994 edition is selling for $6 postpaid.

Midwest Old Threshers Reunion
The real grand-daddy of these shows has to be Midwest Old Threshers, an event that is the highlight of the social calendar for the little (pop. 8,000) town of Mt. Pleasant, Iowa, since 1950. Midwest Old Threshers has become a sort of Mecca for old farm equipment junkies; over 100 operating steam engines, over 300 antique tractors, nearly a thousand antique stationary gasoline engines are on display—along with a few old threshers, corn shellers, and ancient combines. Besides the farm equipment, Midwest Old Threshers includes antique steam trains, trolleys, antique cars and trucks, homemade ice cream, fried chicken, ribs, and more exotic fare. Although over 120,000 people attend during the five days of the event each year, it is conducted quite smoothly and efficiently by the folks in Mt. Pleasant, and the crowds aren't oppressive.
Midwest Old Threshers, 1887 245th Street, Mt. Pleasant, Iowa 52641 (319) 385-8937.

Recommended Reading
Combines; Bill Huxley, Osprey Publishing Ltd.
Full Steam Ahead—J. I. Case Tractors & Equipment 1842 - 1955; Dave Erb and Eldon Brumbaugh, American Society of Agricultural Engineers
The Grain Harvesters, Graeme Quick and Wesley Buchele, American Society of Agricultural Engineers
Threshers; Robert N. Pripps and Andrew Morland, Motorbooks International

Index